Published by: AoPS Incorporated
                   15330 Avenue of Science
                   San Diego, CA 92128
                   info@BeastAcademy.com

ISBN: 978-1-934124-57-4

Beast Academy is a registered trademark of AoPS Incorporated.

Written by Jason Batterson, Shannon Rogers, and Deven Ware
Book Design by Lisa T. Phan
Illustrations by Erich Owen
Grayscales by Greta Selman

Visit the Beast Academy website at BeastAcademy.com.
Visit the Art of Problem Solving website at artofproblemsolving.com.
Printed in the United States of America.
2022 Printing.

Contents:

This is Practice Book 4D in the Beast Academy level 4 series.

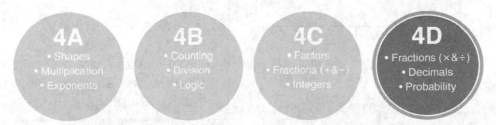

4A
• Shapes
• Multiplication
• Exponents

4B
• Counting
• Division
• Logic

4C
• Factors
• Fractions (+&−)
• Integers

4D
• Fractions (×&÷)
• Decimals
• Probability

For more resources and information, visit BeastAcademy.com/resources.

This is Beast Academy Practice Book 4D.

MATH
PRACTICE
4D

Each chapter of this Practice book corresponds to a chapter from Beast Academy Guide 4D.

MATH
GUIDE
4D

The first page of each chapter includes a recommended sequence for the Guide and Practice books.

You may also read the entire chapter in the Guide before beginning the Practice chapter.

Use this Practice book with Guide 4D from BeastAcademy.com.

Recommended Sequence:

Book	Pages
Guide:	12–25
Practice:	7–16
Guide:	26–31
Practice:	17–27
Guide:	32–37
Practice:	28–37

You may also read the entire chapter in the Guide before beginning the Practice chapter.

Some problems in this book are very challenging. These problems are marked with a ★. The hardest problems have two stars!

Every problem marked with a ★ has a *hint!*

Hints for the starred problems begin on page 102.

Other problems are marked with a ✏. For these problems, you should write an explanation for your answer.

Some pages direct you to related pages from the Guide.

None of the problems in this book require the use of a calculator.

Solutions are in the back, starting on page 106.

A complete explanation is given for every problem!

CHAPTER 10
Fractions

Use this Practice book with
Guide 4D from BeastAcademy.com.

Recommended Sequence:

You may also read the entire chapter
in the Guide before beginning the
Practice chapter.

EXAMPLE | What number is $\frac{1}{3}$ of 15?

We can locate $\frac{1}{3}$ on the number line by splitting the number line between 0 and 1 into three equal pieces. Each piece has length $\frac{1}{3}$.

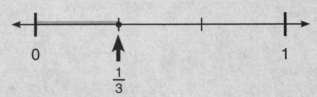

0        $\frac{1}{3}$ 1

We can find a fraction *of* a number!

Similarly, we can locate $\frac{1}{3}$ of 15 on the number line by splitting the number line between 0 and 15 into 3 equal pieces.

The length of each piece is $\frac{1}{3}$ of 15.

The first piece begins at 0 and ends at $\frac{1}{3}$ of 15.

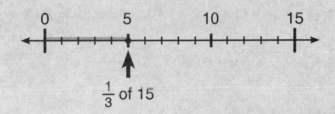

0   5   10   15

$\frac{1}{3}$ of 15

Therefore, $\frac{1}{3}$ of 15 is **5**.

PRACTICE | Answer each question below. You may find the number lines helpful.

1. What number is $\frac{1}{5}$ of 15?

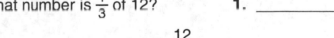

0   15

2. What number is $\frac{1}{3}$ of 12?

0   12

1. _____

2. _____

3. What is $\frac{1}{6}$ of 18?

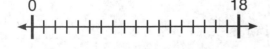

0   18

4. What is $\frac{1}{7}$ of 14?

0   14

3. _____

4. _____

"of"ACTIONS

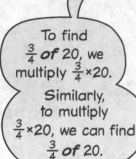

We often use the word "of" to mean multiplication.

To find $\frac{3}{4}$ *of* 20, we multiply $\frac{3}{4} \times 20$.

Similarly, to multiply $\frac{3}{4} \times 20$, we can find $\frac{3}{4}$ *of* 20.

We know that **one** fourth of 20 is 5.

Three fourths of 20 is **three** times as much as **one** fourth of 20.

Therefore, $\frac{3}{4}$ of 20 is $3 \times 5 = \textbf{15}$.

We have
$$\frac{3}{4} \times 20 = \left(3 \times \frac{1}{4}\right) \times 20$$
$$= 3 \times \left(\frac{1}{4} \times 20\right)$$
$$= 3 \times 5$$
$$= \textbf{15}.$$

PRACTICE | Answer each question below.

5. What number is $\frac{2}{9}$ of 36?

6. What number is $\frac{2}{3}$ of 36?

5. _____

6. _____

7. What number is $\frac{3}{5}$ of 15?

8. What number is $\frac{3}{4}$ of 44?

7. _____

8. _____

9. Compute $\frac{7}{9} \times 18$.

10. Compute $\frac{5}{6} \times 18$.

9. _____

10. _____

11. Compute $\frac{5}{17} \times 17$.

12. What number is $\frac{18}{27}$ of 30?

11. _____

12. _____

PRACTICE | Answer each question below.

13. Peter's car holds 16 gallons of gas. Peter uses $\frac{3}{4}$ of a tank of gas to drive to his grandmother's house. How many gallons of gas does Peter use during his drive?

13. _____

14. One foot is 12 inches. How many inches are in $\frac{2}{3}$ of a foot?

14. _____

15. Erica had 6 gallons of paint. She used $\frac{2}{3}$ of her paint to decorate her living room. How many gallons of paint does Erica have left?

15. _____

16. ★ Lisa had $36. She spent $\frac{1}{6}$ of her money on a book. Then, she spent $\frac{1}{5}$ of her *remaining* money on a pencil set. How much money did she spend all together?

16. _____

17. ★ In a 90-minute basketball practice, $\frac{1}{6}$ of the time is spent warming up, $\frac{1}{5}$ of the time is spent doing passing drills, and $\frac{1}{3}$ of the time is spent doing shooting drills. The remaining time is used for a scrimmage. How many minutes long is the scrimmage?

17. _____

EXAMPLE | Reggie jogs $\frac{1}{3}$ of the way to his friend Joey's house and then walks the remaining 4 miles. How far does Reggie travel to get to Joey's house?

We draw a line to represent Reggie's trip.

Reggie jogs $\frac{1}{3}$ of the distance and walks the rest of the way. So, we know that the remaining 4 miles he walks is $\frac{2}{3}$ of the distance.

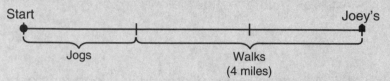

Since $\frac{2}{3}$ of the total distance is 4 miles, each third of the total distance is $4 \div 2 = 2$ miles.

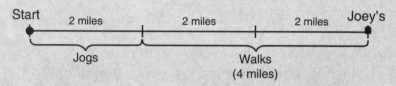

All together, Reggie travels $2 \times 3 = \textbf{6 miles}$.

PRACTICE | Answer each question below.

18. Ralph spent 30 minutes playing soccer this weekend, which was $\frac{1}{7}$ of the time he spent at the park. How many minutes did Ralph spend at the park this weekend?

18. _____

19. When Matt's gas tank is $\frac{1}{5}$ full, it takes an additional 12 gallons of gas to fill the tank. How many gallons does Matt's gas tank hold when full?

19. _____

20. Paul travels by train for the first 600 miles of a trip, then takes a bus for the remaining $\frac{3}{8}$ of the trip. How many miles did Paul travel on his trip?

20. _____

PRACTICE | Answer each question below.

21. Grogg's mom brings a full box of markers to share with the class. After giving the first five students three markers each, she sees that $\frac{1}{4}$ of the box is empty. How many markers were originally in the box?

21. _____

22. After spending $\frac{5}{9}$ of his money on a game, James is left with $8. How many dollars did James spend on his game?

22. _____

23. Lizzie reads for $\frac{5}{6}$ of her study hall time and spends the remaining 15 minutes on math homework. How many minutes long is Lizzie's study hall?

23. _____

24. ★ A middle school math team is made up of students from 6th, 7th, and 8th grade. Five team members are 6th graders, ten are 7th graders, and two sevenths of the team members are 8th graders. How many math team members are 8th graders?

24. _____

25. ★★ Sue runs the first $\frac{7}{10}$ of a race. Then, she jogs $\frac{3}{4}$ of the remaining distance. Finally, she walks the last 375 meters of the race. What is the total length in meters of Sue's race?

25. _____

EXAMPLE | What is $7 \times \frac{3}{5}$?

To compute $7 \times \frac{3}{5}$, we can add 7 copies of $\frac{3}{5}$.

We can multiply any whole number by a fraction.

$$7 \times \frac{3}{5} = \frac{3}{5} + \frac{3}{5} + \frac{3}{5} + \frac{3}{5} + \frac{3}{5} + \frac{3}{5} + \frac{3}{5}$$

$$= \frac{3+3+3+3+3+3+3}{5}$$

$$= \frac{7 \times 3}{5}$$

$$= \frac{21}{5}.$$

Adding 7 copies of $\frac{3}{5}$ gives us $7 \times 3 = 21$ fifths.

So, $7 \times \frac{3}{5} = \frac{7 \times 3}{5} = \frac{21}{5} = 4\frac{1}{5}$.

To multiply a whole number and a fraction, we multiply the whole number by the numerator, and put the result over the denominator:

$$a \times \frac{b}{c} = \frac{a \times b}{c}.$$

PRACTICE | Write each answer in simplest form.

26. $5 \times \frac{1}{12} =$

27. $8 \times \frac{1}{9} =$

28. $4 \times \frac{2}{3} =$

29. $6 \times \frac{5}{7} =$

30. $8 \times \frac{5}{8} =$

31. $8 \times \frac{3}{10} =$

32. $2 \times \frac{7}{5} =$

33. $6 \times \frac{10}{7} =$

34. Alex's snickerdoodle recipe calls for $\frac{2}{3}$ cups of brown sugar for each batch. How many cups of brown sugar does Alex need to make 7 batches of snickerdoodles?

Give your answer as a mixed number in simplest form.

34. _____

EXAMPLE | What is $\frac{7}{11}$ of 13?

To find $\frac{7}{11}$ of 13, we multiply $\frac{7}{11} \times 13$.

Since multiplication is commutative,
we have $\frac{7}{11} \times 13 = 13 \times \frac{7}{11} = \frac{13 \times 7}{11} = \frac{91}{11}$.

So, $\frac{7}{11}$ of 13 is $\frac{91}{11} = 8\frac{3}{11}$.

All four of these expressions are equivalent:

$$\frac{b}{c} \times a = a \times \frac{b}{c} = \frac{a \times b}{c} = \frac{b \times a}{c}.$$

PRACTICE | Write each answer in simplest form.

35. What is $\frac{4}{7}$ of 5?

36. What is $\frac{2}{3}$ of 8?

35. _____

36. _____

37. What is $\frac{5}{8}$ of 3?

38. What is $\frac{6}{11}$ of 4?

37. _____

38. _____

39. Compute $\frac{2}{9} \times 4$.

40. Compute $\frac{5}{8} \times 7$.

39. _____

40. _____

41. What is the perimeter of a square field whose sides are $\frac{2}{5}$ miles long? *Give your answer as a mixed number in simplest form and include units.*

41. _____

42. Which weighs more: $\frac{4}{9}$ of a 7-pound bag of flour, or seven $\frac{4}{9}$-pound bags of flour?

FRACTIONS
Multiplying Fractions

Rewriting an expression can make a computation easier.

EXAMPLE | Compute $21 \times \frac{5}{7}$.

$21 \times \frac{5}{7} = \frac{21 \times 5}{7} = \frac{105}{7}$. Then, $105 \div 7 = 15$. So, $21 \times \frac{5}{7} = \mathbf{15}$.

— *or* —

We notice that 21 is a multiple of 7, so we can make some of our computations easier by rewriting the expression as shown:

$$21 \times \frac{5}{7} = \frac{21 \times 5}{7}$$
$$= \frac{21}{7} \times 5$$
$$= 3 \times 5$$
$$= \mathbf{15}.$$

In general, we have
$$a \times \frac{b}{c} = \frac{a \times b}{c} = \frac{a}{c} \times b.$$

PRACTICE | Write each answer in simplest form.

For an extra challenge, see how many products you can compute without writing anything down!

43. $\frac{4}{11} \times 55 =$

44. $32 \times \frac{3}{8} =$

45. $13 \times \frac{17}{39} =$

46. $99 \times \frac{7}{9} =$

47. $6 \times \frac{10}{15} =$

48. $12 \times \frac{10}{3} =$

49. Compute $\left(66 \times \frac{5}{11}\right) + \left(\frac{4}{13} \times 26\right)$.

49. _____

50. ★ Compute $\frac{8}{9} \times \left(\frac{3}{4} \times 54\right)$.

50. _____

EXAMPLE | What fraction of 20 is 12?

We consider a bag of 20 blocks.
If 12 of the 20 blocks are blue, then
$\frac{12}{20}$ of the blocks are blue.
We simplify $\frac{12}{20}$ to $\frac{3}{5}$. So, 12 is $\frac{3}{5}$ of 20.

We check that $\frac{3}{5}$ of 20 is $\frac{3}{5} \times 20 = 12$. ✔

PRACTICE | Write each answer as a fraction in simplest form.

51. What fraction of 44 is 16?

51. _____

52. 14 is what fraction of 63?

52. _____

53. What fraction of 40 is 31?

53. _____

54. Ted's homework has 22 questions. Ted completed 12 of the questions last night and finished the rest this morning. What fraction of his homework questions did Ted answer last night?

54. _____

55. Alice has 27 toy cars. Six of those cars are missing a wheel. What fraction of Alice's toy cars are *not* missing a wheel?

55. _____

56. Twenty students in Nick's basketweaving class have brown eyes. The other five students have blue eyes. What fraction of the students have brown eyes?

56. _____

Multiplying Fractions

EXAMPLE | Fill in the blank with the correct number in simplest form: $\square \times 14 = 30$.

We use the relationship between multiplication and division:

If $\square \times 14 = 30$, then $30 \div 14 = \square$.

So, the number we are looking for is $30 \div 14 = \frac{30}{14}$, which can be simplified to $\frac{15}{7}$, or $2\frac{1}{7}$.

We check that $\boxed{\frac{15}{7}} \times 14 = 30$. ✓

PRACTICE | Fill in the blank in each equation below with the correct number in simplest form.

57. $\boxed{} \times 8 = 54$.

58. $15 \times \boxed{} = 24$.

59. $21 \times \boxed{} = 9$.

60. $\boxed{} \times 12 = 20$.

PRACTICE | Answer each question below in simplest form.

61. If $a \times 26 = 30$, what is the value of a?

61. _____

62. If $60 \times b = 35$, what is the value of b?

62. _____

63. 24 is what fraction of 20?
★

63. _____

PRACTICE | Answer each question below in simplest form, using mixed numbers when possible.

64. What is the result when $\frac{3}{5}$ of 16 is added to $\frac{2}{5}$ of 10?

64. _____

65. Compute $\left(\frac{1}{6}\times 29\right)+\left(\frac{1}{6}\times 31\right)$.

65. _____

66. What number is $\frac{5}{3}$ of the product of 23 and 6?

66. _____

67. Draw a line to mark the top of the water when the 2-cup measuring glass below is $\frac{5}{8}$ full.

68. Bobby Bug walks from his house to his best friend Sammy Slug's house along the straight path shown below. He stops to rest $\frac{7}{10}$ of the way there. How far does Bobby walk before he stops to rest? Mark your answer with an arrow on the ruler below.

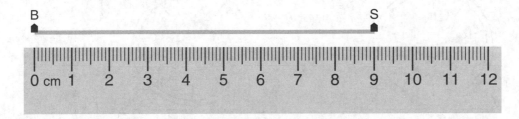

PRACTICE | Answer each question below in simplest form, using mixed numbers when possible.

69. Two paperclips are placed end-to-end as shown below. Draw an arrow to mark the length of five paperclips placed end-to-end on the ruler below.

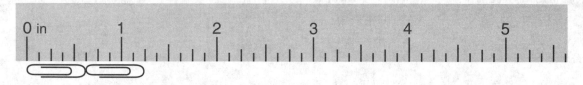

70. Together, 2 identical blocks weigh $\frac{7}{8}$ of a pound. Draw an arrow to display the correct weight of 8 identical blocks on the scale on the right.

 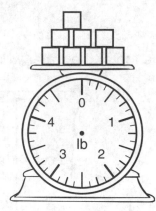

71. Draw an arrow to display the correct weight when $\frac{3}{8}$ of the pie has been eaten.

PRACTICE | Answer each question below.

72. Find a simpler expression that equals $\frac{n}{20} \times 100$.

72. _____

73. Winnie gives $\frac{2}{3}$ of her 24 peanuts to Lizzie. Lizzie gives $\frac{1}{4}$ of the peanuts she gets from Winnie to Alex. Alex gives $\frac{1}{2}$ of these peanuts to Grogg. How many peanuts does Grogg get from Alex?

73. _____

74. There are 48 students enrolled in Sam's Saturday karate class. Last Saturday, $\frac{3}{4}$ of the students were in class. This Saturday, $\frac{1}{6}$ of the students **missed** class. How many more students were in class this Saturday than last Saturday?

74. _____

75. What number is $\frac{1}{3}$ of $\frac{1}{2}$ of $\frac{1}{5}$ of 900?

75. _____

76. If $\frac{5}{12}$ of n is 10, what is $\frac{7}{12}$ of n?

76. _____

PRACTICE | Answer each question below. You may assume all variables are nonzero.

77. $\frac{4}{11}$ of m equals 12. What is the value of m?

77. $m =$ _____

78. Find a simpler expression that equals $\frac{v}{w} \times w$.

78. _____

79. ★ ✏️ Is $\frac{1}{a}$ of $\frac{1}{b}$ of 100 equal to $\frac{1}{b}$ of $\frac{1}{a}$ of 100? Explain.

80. $\frac{1}{4}$ of $\frac{1}{5}$ of 100 is equal to $\frac{1}{n}$ of 100. What is the value of n?

80. $n =$ _____

81. ★ ★ $\frac{1}{2}$ of $\frac{1}{3}$ of y is the same as $\frac{1}{z}$ of y. What is the value of z?

81. $z =$ _____

PRACTICE

For each problem below, use the given digits to fill in the blanks so that the value of the expression is *as large as possible*. Compute the value of the expression you create as a mixed number in simplest form.

82. Digits: 2, 3, 5

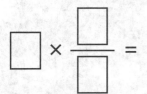

83. Digits: 4, 6, 7

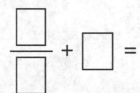

84. ★ Digits: 3, 4, 7, 8

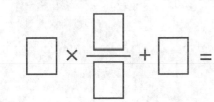

85. ★ Digits: 2, 3, 5, 7

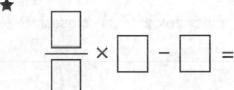

86. ★ Digits: 4, 5, 6, 7

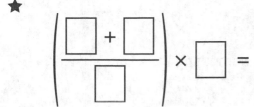

87. ★ Digits: 3, 5, 8, 9

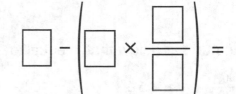

EXAMPLE | Estimate $\frac{29}{41} \times 205$.

$\frac{29}{41}$ is close to $\frac{30}{40} = \frac{3}{4}$, and 205 is close to 200.

So, we expect $\frac{29}{41}$ of 205 to be close to $\frac{3}{4} \times 200$.

$\frac{3}{4} \times 200 = 3 \times \frac{200}{4} = 3 \times 50 = 150$, so we estimate that

$$\frac{29}{41} \times 205 \approx \frac{3}{4} \times 200 = \mathbf{150}.$$

In fact, $\frac{29}{41} \times 205 = 145$. We overestimated, and our estimate differs from the actual value by $150 - 145 = 5$.

When compared to 150, this difference of 5 is pretty small. So, 150 is a good estimate of $\frac{29}{41} \times 205$.

> A good estimate is easy to compute **and** is close to the actual answer.

PRACTICE | Solve each problem below by estimating.

88. Circle the number below that is closest to $\frac{31}{210}$.

$\frac{1}{700}$ $\qquad$ $\frac{1}{70}$ $\qquad$ $\frac{1}{7}$ $\qquad$ 1

89. Circle the number below that is closest to $\frac{25}{82}$.

$\frac{1}{3}$ $\qquad$ $\frac{2}{3}$ $\qquad$ 1 $\qquad$ $\frac{5}{3}$

90. Circle the number below that $\frac{6}{55}$ of 70 is closest to.

2 $\qquad$ 8 $\qquad$ 15 $\qquad$ 22

91. Circle the number below that is equal to $\frac{58}{79}$ of 24 without computing the exact answer.

$2\frac{49}{79}$ $\qquad$ $5\frac{49}{79}$ $\qquad$ $10\frac{49}{79}$ $\qquad$ $17\frac{49}{79}$ $\qquad$ $28\frac{49}{79}$

We can use estimation to make sure our answers are reasonable.

EXAMPLE | Janet computes $\frac{19}{83} \times 42$.

Is her answer of $20\frac{2}{83}$ reasonable?

$\frac{19}{83}$ is close to $\frac{20}{80} = \frac{1}{4}$ and 42 is close to 40.

So, we estimate that $\frac{19}{83} \times 42$ is close to $\frac{1}{4} \times 40 = 10$.

Janet's answer is more than 20, which is more than twice our estimate. So, Janet's answer of $20\frac{2}{83}$ is **not reasonable**.

In fact, $\frac{19}{83} \times 42 = 9\frac{51}{83}$.

An answer is **reasonable** if it makes sense.

By using estimation to check whether our answers are reasonable or not, we can often catch mistakes and correct them.

PRACTICE | Use estimation to answer each question below.

92. Circle the number below that is closest to $\frac{21}{51} \times 12{,}546$.

1,000   5,000   20,000   400,000

93. Circle the number below that is closest to $\frac{29}{21} \times 13{,}982$.

1,000   5,000   20,000   400,000

94. $\frac{6}{19} \times 58$ is between which two consecutive multiples of 10?

94. Between _____ and _____

95. $298 \times \frac{25}{109}$ is between which two consecutive multiples of 20?

95. Between _____ and _____

96. Without computing the products, order these expressions from least to greatest.

A. $\frac{41}{104} \times 13$ **B.** $7 \times \frac{35}{37}$ **C.** $4 \times \frac{5}{26}$ **D.** $\frac{3}{28} \times 33$

_____ < _____ < _____ < _____

EXAMPLE | Compute $10 \times 4\frac{2}{3}$.

We can convert $4\frac{2}{3}$ into a fraction: $4\frac{2}{3} = \frac{14}{3}$. Then, we multiply:

$$10 \times 4\frac{2}{3} = 10 \times \frac{14}{3}$$
$$= \frac{10 \times 14}{3}$$
$$= \frac{140}{3}.$$

We write $\frac{140}{3}$ as a mixed number: $\frac{140}{3} = 46\frac{2}{3}$.

So, $10 \times 4\frac{2}{3} = \mathbf{46\frac{2}{3}}$.

$-\ or\ -$

We use the distributive property. $4\frac{2}{3}$ equals $4 + \frac{2}{3}$, so $10 \times 4\frac{2}{3}$ equals $10 \times \left(4 + \frac{2}{3}\right)$. We distribute the 10 and get

$$10 \times 4\frac{2}{3} = 10 \times \left(4 + \frac{2}{3}\right)$$
$$= \left(10 \times 4\right) + \left(10 \times \frac{2}{3}\right)$$
$$= 40 + \frac{20}{3}$$
$$= 40 + 6\frac{2}{3}$$
$$= 46\frac{2}{3}.$$

So, $10 \times 4\frac{2}{3} = \mathbf{46\frac{2}{3}}$.

PRACTICE | Write each product below as a whole or mixed number in simplest form.

97. $7 \times 1\frac{1}{4} =$

98. $15 \times 2\frac{1}{3} =$

99. $5\frac{3}{4} \times 9 =$

100. $8\frac{5}{6} \times 11 =$

101. $9 \times 6\frac{1}{11} =$

102. $3\frac{4}{7} \times 12 =$

PRACTICE	Answer each question below. Write each answer as a whole or mixed number in simplest form.

103. Wilson is $33\frac{3}{4}$ inches tall. Wilson's dad is twice as tall as Wilson. How many inches tall is Wilson's dad?

103. _____

104. A dodecagon is a polygon with 12 sides. How many inches long is the perimeter of a regular dodecagon with side length $5\frac{3}{8}$ inches?

104. _____

105. To prepare for the Beast Academy fitness test, Alex runs $1\frac{3}{4}$ miles twice per week for four weeks. How many total miles does Alex run in those four weeks?

105. _____

106. Three identical bags of ice weigh a total of 8 pounds. How many pounds do ten identical bags of ice weigh together?

106. _____

107. ★ Amber is painting a rectangular wall that is 15 feet tall and $18\frac{2}{3}$ feet wide. She uses 1 quart of paint for every 75 square feet of wall. How many quarts of paint will she use to paint the entire wall?

107. _____

PRACTICE | Compute each area or perimeter below.
Remember to include units where necessary.

108. What is the perimeter of the regular pentagon shown below?

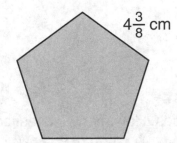

$4\frac{3}{8}$ cm

108.

109. What is the area of the rectangle shown below?

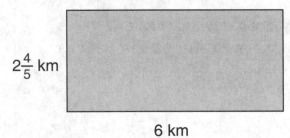

$2\frac{4}{5}$ km

6 km

109.

110. Four congruent right triangles are attached as shown below. What is the area of the shaded region formed by the triangles?

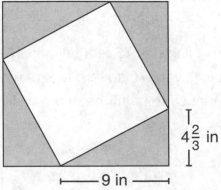

$4\frac{2}{3}$ in

9 in

110.

PRACTICE | Compute each area or perimeter below.
Remember to include units where necessary.

---

111. A rectangle and an isosceles triangle with the same height
are attached as shown below to make a pentagon. What is
the area of the pentagon?

111. _____

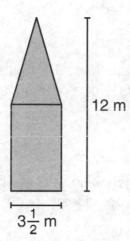

12 m

$3\frac{1}{2}$ m

112. Three congruent rectangles are attached as shown below to create
a large rectangle. What is the area of the large rectangle?

112. _____

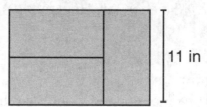

11 in

113. Five equilateral triangles are arranged as shown to make a
quadrilateral. The perimeter of each triangle is 5 feet. What is the
perimeter of the quadrilateral?

113. _____

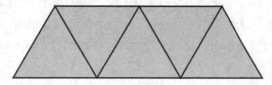

EXAMPLE

Eight pounds of flour are divided into $\frac{1}{5}$-pound bags. How many bags are needed to hold all the flour?

From each pound of flour, we can make five $\frac{1}{5}$-pound bags, so from eight pounds of flour, we can make $8 \times 5 = \textbf{40}$ bags.

PRACTICE | Answer each question below.

114. The Burger Palace orders a box that contains 12 pounds of hamburger patties. Each patty weighs $\frac{1}{3}$ of a pound. How many patties are in the box?

114. _____

115. One lap around the track at Kayla's school is $\frac{1}{5}$ of a mile. Kayla runs 7 miles around the track. How many laps does Kayla run?

115. _____

116. Captain Kraken has sheets of plywood that are $\frac{1}{4}$ inches thick. A stack of these plywood sheets is 18 inches tall. How many sheets are in the stack?

116. _____

117. A regular polygon whose sides are $\frac{1}{5}$ of an inch long has a perimeter of 4 inches. How many sides does the polygon have?

117. _____

118. The Beast Island Candy Shop makes 50 pounds of peppermint bark candy. The candy makers split the bark into $\frac{1}{8}$-pound bags, which are each sold for $6. How much money will the shop collect if they sell every bag of peppermint bark?

118. _____

The **_reciprocal_** of a number n is the number we multiply n by to get 1.

EXAMPLE | What is the reciprocal of 7?

$7 \times \frac{1}{7} = 1$, so 7 and $\frac{1}{7}$ are reciprocals.

$\frac{1}{7}$ is the reciprocal of 7.

PRACTICE | Find the reciprocal of each number or expression below.

119. $\frac{1}{5}$ Reciprocal: _____

120. 17   Reciprocal: _____

121. $5+6$ Reciprocal: _____

122. 6×9 Reciprocal: _____

123. $\frac{15}{3}$ Reciprocal: _____

124. $\frac{13}{39}$ Reciprocal: _____

125. Does 0 have a reciprocal? If so, what is it? If not, explain why not.

126. Write an expression for the reciprocal of n. (Assume that n is not zero.)

126. _____

127. Write an expression for the reciprocal of $\frac{1}{a}$. (Assume that a is not zero.)

127. _____

128. Write an expression for the reciprocal of $c+1$. (Assume that c is not -1.)

128. _____

129. What is the sum of the reciprocals of $\frac{1}{5}$ and $\frac{1}{7}$?

129. _____

EXAMPLE | What is $8 \div \frac{1}{5}$?

We consider dividing 8 pounds of flour into $\frac{1}{5}$-pound bags. From each pound of flour, we can make five $\frac{1}{5}$-pound bags, so from eight pounds of flour, we can make $8 \times 5 = \textbf{40}$ bags.

— *or* —

We look at the number line to find out how many $\frac{1}{5}$'s are in 8. Since there are 5 fifths in 1, there are $8 \times 5 = \textbf{40}$ fifths in 8.

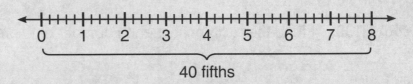

40 fifths

We write $8 \div \frac{1}{5} = 8 \times 5 = \textbf{40}$.

Dividing by a number is the same as multiplying by its reciprocal.

Dividing by n is the same as multiplying by the *reciprocal* of n.

For example, $9 \div 4 = 9 \times \frac{1}{4} = \frac{9}{4}$, and $5 \div \frac{1}{7} = 5 \times 7 = 35$.

PRACTICE | To compute each quotient below, multiply by the reciprocal. Write your answers in simplest form.

130. $5 \div \frac{1}{7} =$

131. $3 \div \frac{1}{16} =$

132. $9 \div \frac{1}{4} =$

133. $\frac{1}{16} \div \frac{1}{8} =$

PRACTICE | Write each quotient below in simplest form.

134. $3\frac{2}{11} \div \frac{1}{2} =$

135. $2\frac{1}{5} \div \frac{1}{8} =$

136. $5 \div \left(3 \div \frac{1}{12}\right) =$

137. $(5 \div 3) \div \frac{1}{12} =$

138. $\left(9 \div \frac{1}{10}\right) \div \frac{1}{5} =$

139. $9 \div \left(\frac{1}{10} \div \frac{1}{5}\right) =$

PRACTICE | Answer each question below.

140. How many $\frac{1}{4}$-cup scoops of flour are needed to equal $2\frac{3}{4}$ cups?

140. _____

141. The tallest tree in Ranger Rick's forest grows $\frac{1}{8}$ of an inch every week. How many weeks will it take for the tree to grow $7\frac{3}{4}$ inches?

141. _____

142. Tara brought 4 gallons of water on a hike. She gave $\frac{1}{3}$ of a gallon to each of her hiking companions, which left her with $1\frac{1}{3}$ gallons of water. How many companions were on the hike with Tara?

142. _____

143. If $a \div \frac{1}{16} = 20$, what is the value of a?
★

143. _____

EXAMPLE | Fill in the missing entries in the Cross-Number puzzle below to make all the equations true.

	×	$\frac{1}{5}$	=	$\frac{4}{5}$
×		×		×
$2\frac{1}{3}$	×	15	=	
=		=		=
	×		=	

We fill in the missing entries as shown below.

$\boxed{4} \times \frac{1}{5} = \frac{4}{5}$.　　　　$4 \times 2\frac{1}{3} = \boxed{9\frac{1}{3}}$.　　　　$9\frac{1}{3} \times 3 = \boxed{28}$, **or**

$2\frac{1}{3} \times 15 = \boxed{35}$.　　　　$\frac{1}{5} \times 15 = \boxed{3}$.　　　　$\frac{4}{5} \times 35 = \boxed{28}$.

4	×	$\frac{1}{5}$	=	$\frac{4}{5}$
×		×		×
$2\frac{1}{3}$	×	15	=	**35**
=		=		=
	×		=	

4	×	$\frac{1}{5}$	=	$\frac{4}{5}$
×		×		×
$2\frac{1}{3}$	×	15	=	**35**
=		=		=
$9\frac{1}{3}$	×	**3**	=	

4	×	$\frac{1}{5}$	=	$\frac{4}{5}$
×		×		×
$2\frac{1}{3}$	×	15	=	**35**
=		=		=
$9\frac{1}{3}$	×	**3**	=	**28**

You can find more of these puzzles
at BeastAcademy.com!

PRACTICE | Fill in the missing entries in the Cross-Number puzzles below to make all the equations true. Each entry should be in simplest form. Use whole and mixed numbers where possible.

144.

6	×	$\frac{2}{5}$	=	
×		×		×
$1\frac{1}{3}$	×	18	=	
=		=		=
	×		=	

145.

60	×		=	2
÷		÷		×
		÷		=	30
=		=		=
10	÷		=	

PRACTICE | Fill in the missing entries in the Cross-Number puzzles below to make all the equations true.

146.

3	×		=	12
÷		÷		÷
$\frac{1}{35}$	×	7	=	
=		=		=
	×		=	

147.

18	×		=	30
×		÷		×
$\frac{1}{4}$	÷	$\frac{1}{9}$	=	
=		=		=
	×		=	

148.

35	×		=	15
×		÷		×
	÷	$\frac{1}{4}$	=	
=		=		=
14	×		=	

149. ★

	×	5	=	
÷		÷		÷
$\frac{1}{36}$	×	4	=	
=		=		=
24	×		=	

In a **Fraction Fill** puzzle, each cell is labeled with a fractional **clue**. The goal is to shade some of the cells in the grid so that each clue gives the fraction of a cell's **neighborhood** cells (surrounding cells, including the cell containing the clue) that are shaded.

A **corner clue** gives the fraction of its 4 neighborhood cells that are shaded.

A **side clue** gives the fraction of its 6 neighborhood cells that are shaded.

A **middle clue** gives the fraction of its 9 neighborhood cells that are shaded.

Below is a completed fraction fill puzzle.

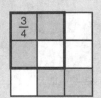

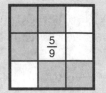

3/4	1/2	1/4
2/3	5/9	1/2
1/2	1/2	1/2

EXAMPLE | Solve the Fraction Fill puzzle to the right.

3/4	2/3	2/3	1/2
5/6	2/3	5/9	1/3
1	2/3	1/3	0

We begin by looking for clues that help us immediately.

The 1 in the bottom-left corner tells us that $1 = \frac{4}{4}$ of its neighborhood cells are shaded. So, we shade all four cells in the bottom-left corner.

The 0 in the bottom-right corner tells us that $0 = \frac{0}{4}$ of its neighborhood cells are shaded.

We circle these four cells to remind ourselves that they should not be shaded.

The $\frac{5}{9}$ indicated in the second row tells us that $\frac{5}{9}$ of its neighborhood cells are shaded. Previously, we shaded two of the nine cells. We also found that four of these nine cells cannot be shaded. So, the three neighborhood cells that remain must all be shaded.

The $\frac{3}{4}$ in the top-left corner tells us that $\frac{3}{4}$ of its neighborhood cells are shaded. Since we have already shaded 3 of these 4 cells, the fourth cell must be left unshaded.

Every cell in the grid has been shaded or circled (to tell us it is unshaded), so we are done!

PRACTICE | Solve each Fraction Fill puzzle below.

150.

$\frac{3}{4}$	$\frac{2}{3}$	$\frac{3}{4}$
$\frac{2}{3}$	$\frac{2}{3}$	$\frac{5}{6}$
$\frac{1}{2}$	$\frac{2}{3}$	1

151.

$\frac{3}{4}$	$\frac{5}{6}$	1
$\frac{1}{2}$	$\frac{5}{9}$	$\frac{2}{3}$
$\frac{1}{2}$	$\frac{1}{2}$	$\frac{1}{2}$

152.

$\frac{3}{4}$	$\frac{2}{3}$	$\frac{2}{3}$	$\frac{1}{2}$
$\frac{1}{2}$	$\frac{5}{9}$	$\frac{2}{3}$	$\frac{2}{3}$
$\frac{1}{4}$	$\frac{1}{2}$	$\frac{5}{6}$	1

153.

$\frac{1}{2}$	$\frac{2}{3}$	$\frac{2}{3}$	$\frac{3}{4}$
$\frac{1}{3}$	$\frac{5}{9}$	$\frac{5}{9}$	$\frac{2}{3}$
0	$\frac{1}{3}$	$\frac{1}{2}$	$\frac{3}{4}$

154.

0	$\frac{1}{3}$	$\frac{1}{3}$	$\frac{1}{2}$
$\frac{1}{6}$	$\frac{4}{9}$	$\frac{1}{3}$	$\frac{1}{2}$
$\frac{1}{4}$	$\frac{1}{2}$	$\frac{1}{3}$	$\frac{1}{2}$

155.

$\frac{3}{4}$	$\frac{2}{3}$	$\frac{2}{3}$	$\frac{3}{4}$
$\frac{5}{6}$	$\frac{2}{3}$	$\frac{2}{3}$	$\frac{2}{3}$
1	$\frac{2}{3}$	$\frac{2}{3}$	$\frac{1}{2}$

156.

$\frac{1}{4}$	$\frac{1}{6}$	0	$\frac{1}{3}$	$\frac{1}{2}$
$\frac{1}{2}$	$\frac{4}{9}$	$\frac{1}{3}$	$\frac{4}{9}$	$\frac{1}{2}$
$\frac{1}{3}$	$\frac{1}{3}$	$\frac{4}{9}$	$\frac{5}{9}$	$\frac{2}{3}$
$\frac{1}{2}$	$\frac{1}{2}$	$\frac{2}{3}$	$\frac{2}{3}$	$\frac{3}{4}$

157.

1	1	1	$\frac{2}{3}$	$\frac{1}{2}$
$\frac{2}{3}$	$\frac{7}{9}$	$\frac{8}{9}$	$\frac{2}{3}$	$\frac{1}{2}$
$\frac{1}{2}$	$\frac{5}{9}$	$\frac{7}{9}$	$\frac{2}{3}$	$\frac{2}{3}$
$\frac{1}{4}$	$\frac{1}{3}$	$\frac{2}{3}$	$\frac{2}{3}$	$\frac{3}{4}$

Try these!

PRACTICE | Solve each Fraction Fill puzzle below.

158.

$\frac{3}{4}$	$\frac{5}{6}$	$\frac{5}{6}$	$\frac{5}{6}$	$\frac{2}{3}$	$\frac{3}{4}$
$\frac{2}{3}$	$\frac{2}{3}$	$\frac{2}{3}$	$\frac{5}{9}$	$\frac{4}{9}$	$\frac{1}{2}$
$\frac{1}{2}$	$\frac{4}{9}$	$\frac{5}{9}$	$\frac{1}{3}$	$\frac{1}{3}$	$\frac{1}{3}$
$\frac{1}{2}$	$\frac{1}{3}$	$\frac{1}{3}$	0	0	0

159.

$\frac{1}{2}$	$\frac{1}{3}$	0	$\frac{1}{3}$	$\frac{1}{2}$	$\frac{3}{4}$
$\frac{1}{2}$	$\frac{4}{9}$	$\frac{1}{3}$	$\frac{5}{9}$	$\frac{2}{3}$	$\frac{5}{6}$
$\frac{2}{3}$	$\frac{2}{3}$	$\frac{5}{9}$	$\frac{5}{9}$	$\frac{2}{3}$	$\frac{5}{6}$
$\frac{3}{4}$	$\frac{5}{6}$	$\frac{5}{6}$	$\frac{2}{3}$	$\frac{2}{3}$	$\frac{3}{4}$

160.

$\frac{1}{2}$	$\frac{1}{3}$	$\frac{1}{2}$	$\frac{1}{3}$	$\frac{1}{3}$	0
$\frac{2}{3}$	$\frac{5}{9}$	$\frac{2}{3}$	$\frac{4}{9}$	$\frac{1}{3}$	0
1	$\frac{7}{9}$	$\frac{2}{3}$	$\frac{4}{9}$	$\frac{4}{9}$	$\frac{1}{3}$
$\frac{5}{6}$	$\frac{7}{9}$	$\frac{7}{9}$	$\frac{2}{3}$	$\frac{2}{3}$	$\frac{2}{3}$
$\frac{2}{3}$	$\frac{2}{3}$	$\frac{7}{9}$	$\frac{2}{3}$	$\frac{7}{9}$	$\frac{5}{6}$
$\frac{1}{2}$	$\frac{2}{3}$	1	$\frac{5}{6}$	$\frac{5}{6}$	$\frac{3}{4}$

161.

$\frac{3}{4}$	$\frac{2}{3}$	$\frac{1}{2}$	$\frac{1}{2}$	$\frac{1}{2}$	$\frac{1}{2}$
$\frac{2}{3}$	$\frac{5}{9}$	$\frac{4}{9}$	$\frac{1}{3}$	$\frac{4}{9}$	$\frac{1}{2}$
$\frac{5}{6}$	$\frac{2}{3}$	$\frac{5}{9}$	$\frac{1}{3}$	$\frac{4}{9}$	$\frac{1}{2}$
$\frac{2}{3}$	$\frac{4}{9}$	$\frac{2}{9}$	0	$\frac{1}{3}$	$\frac{1}{2}$
$\frac{5}{6}$	$\frac{2}{3}$	$\frac{4}{9}$	$\frac{2}{9}$	$\frac{1}{3}$	$\frac{1}{3}$
$\frac{3}{4}$	$\frac{2}{3}$	$\frac{1}{2}$	$\frac{1}{3}$	$\frac{1}{3}$	$\frac{1}{4}$

PRACTICE | Write each answer in simplest form.

162. 50 is the product of Kyle's favorite number and 35. What is Kyle's favorite number?

162. _____

163. ★ What number divided by $\frac{1}{5}$ gives the quotient 70?

163. _____

164. ★ Laura's chocolate cake recipe calls for 3 eggs, $\frac{1}{3}$ cups of cocoa powder, $\frac{1}{2}$ cups of sugar, and $\frac{1}{4}$ cups of butter. If Laura has only the ingredients below, how many whole chocolate cakes can she make?

164. _____

- 50 eggs
- 6 cups of cocoa powder
- 7 cups of sugar
- 8 cups of butter

165. ★ When Billy adds $2\frac{1}{2}$ to $\frac{4}{5}$ of his favorite number, the result is his favorite number. What is Billy's favorite number?

165. _____

166. ✏ When we divide one whole number by another, our answer is not always a whole number. For example, $15 \div 7 = 2\frac{1}{7}$. When we divide a whole number by a unit fraction, will our answer **always** be a whole number? If it will, explain why. If it will not, give an example.

37

CHAPTER 11
Decimals

Use this Practice book with
Guide 4D from BeastAcademy.com.

Recommended Sequence:

Book	Pages:
Guide:	38-49
Practice:	39-49
Guide:	50-61
Practice:	50-61
Guide:	62-69
Practice:	62-71

You may also read the entire chapter
in the Guide before beginning the
Practice chapter.

Decimal numbers provide another way to write fractions.

Every decimal number includes a **decimal point** that separates the whole number part from the fractional part of the decimal number.

In a decimal number, each digit has a place value. Each place value to the right of the decimal point is a unit fraction whose denominator is a power of 10:

$$\frac{2}{100's} \quad \frac{3}{10's} \quad \frac{4}{1's} \ . \ \frac{5}{\frac{1}{10}'s} \quad \frac{6}{\frac{1}{100}'s} \quad \frac{7}{\frac{1}{1,000}'s} \quad \frac{8}{\frac{1}{10,000}'s}$$

For example, 4.7 has a 4 in the ones place and a 7 in the tenths place. So, $4.7 = 4\frac{7}{10}$.

$$\underline{4} \ . \ \underline{7}$$

We can read 4.7 as "four point seven."

0.09 has 0 ones, 0 tenths, and 9 hundredths. So, $0.09 = \frac{9}{100}$.

$$\underline{0} \ . \ \underline{0} \ \underline{9}$$

0.09 is read "zero point zero nine."

0.003 has 0 ones, 0 tenths, 0 hundredths, and 3 thousandths. So, $0.003 = \frac{3}{1,000}$.

$$\underline{0} \ . \ \underline{0} \ \underline{0} \ \underline{3}$$

PRACTICE | Write each decimal number below as a fraction with a denominator that is a power of 10.

1. 0.7 = _____

2. 0.3 = _____

3. 0.04 = _____

4. 0.05 = _____

5. 0.01 = _____

6. 0.008 = _____

7. 0.009 = _____

8. 0.006 = _____

9. 0.08 = _____

10. 0.0003 = _____

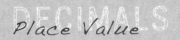

EXAMPLE | Place a decimal point between two of the digits below so that the resulting decimal number has a 4 in the tenths place.

23456

The tenths place is the first place value to the right of the decimal point. To create a decimal number with a 4 in the tenths place, we place the decimal point to the left of the 4:

23.456

PRACTICE | Complete each of the following problems.

11. Circle the hundredths digit of each decimal number below:

945.86   165.402   82.945

12. Place a decimal point between two digits of each number below so that each resulting number has a 5 in the hundredths place.

3045   1459   53501

13. Place a decimal point between two digits of each number below so that each resulting number has tens digit 3 and tenths digit 4.

3242   34434   34334

14. Place a decimal point between two digits of each number below so that each resulting number has the same digit in the hundreds and tenths places.

78768   11221   505404

15. Place a decimal point between two digits of each number below so that each resulting number has a tenths digit that is less than its thousandths digit.

15324   86876   92024

Place Value

EXAMPLE | Fill in the missing numerator and denominator to make the equation below true.

$$0.3 = \frac{3}{} = \frac{}{100}$$

The decimal 0.3 has a 3 in the tenths place. So, $0.3 = \frac{3}{10}$. Then, to convert $\frac{3}{10}$ into an equivalent fraction with denominator 100, we multiply both its numerator and denominator by 10.

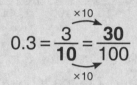

$$0.3 = \frac{3}{10} = \frac{30}{100}$$

We usually just say "decimal" to refer to a decimal number.

PRACTICE | Fill in the missing numerators and denominators to make each equation below true.

16. $0.7 = \dfrac{}{10} = \dfrac{}{1,000}$

17. $0.02 = \dfrac{2}{} = \dfrac{}{1,000}$

18. $0.09 = \dfrac{}{100} = \dfrac{}{1,000}$

19. $0.5 = \dfrac{}{100} = \dfrac{}{1,000}$

20. $0.600 = \dfrac{}{10} = \dfrac{}{1,000}$

21. $0.08 = \dfrac{8}{} = \dfrac{}{1,000}$

22. $0.040 = \dfrac{}{100} = \dfrac{}{1,000}$

23. $0.1 = \dfrac{}{10} = \dfrac{}{100}$

24. Write a decimal that is equal to $\frac{30}{1,000}$.

24. _____

EXAMPLE | Write $\frac{387}{1,000}$ as a decimal.

To begin, we write $\frac{387}{1,000}$ as $\frac{300}{1,000}+\frac{80}{1,000}+\frac{7}{1,000}$.

Next, we simplify $\frac{300}{1,000}$ to $\frac{3}{10}$, and we simplify $\frac{80}{1,000}$ to $\frac{8}{100}$.

So, we have:

$$\frac{387}{1,000} = \frac{300}{1,000}+\frac{80}{1,000}+\frac{7}{1,000}$$

$$= \frac{3}{10} + \frac{8}{100} +\frac{7}{1,000}.$$

To write $\frac{387}{1,000}$ as a decimal, we write 3 in the tenths place, 8 in the hundredths place, and 7 in the thousandths place.

$$\underline{\underset{1's}{0}} \bullet \underset{\frac{1}{10}'s}{3}\ \underset{\frac{1}{100}'s}{8}\ \underset{\frac{1}{1,000}'s}{7}$$

We write a zero in the ones place to make the decimal easier to read.

So, $\frac{387}{1,000} = \mathbf{0.387}$.

PRACTICE | Fill in the missing numerators in each equation below. Then, write each fraction or mixed number as a decimal.

25. $\dfrac{27}{100} = \dfrac{20}{100}+\dfrac{}{100} = \dfrac{}{10}+\dfrac{}{100} = $ ___.___ ___

26. $\dfrac{291}{1,000} = \dfrac{}{1,000}+\dfrac{90}{1,000}+\dfrac{}{1,000} = \dfrac{}{10}+\dfrac{}{100}+\dfrac{}{1,000} = $ ___.___ ___ ___

27. $7\dfrac{54}{100} = 7+\dfrac{}{100}+\dfrac{}{100} = 7+\dfrac{}{10}+\dfrac{}{100} = $ ___.___ ___

28. $\dfrac{56}{1,000} = $ ___.___ ___ ___

EXAMPLE | Write 0.493 as a fraction.

To begin, we write 0.493 as $\frac{4}{10}+\frac{9}{100}+\frac{3}{1,000}$.

In order to add, we convert each of $\frac{4}{10}$ and $\frac{9}{100}$ into equivalent fractions with denomimator 1,000: $\frac{4}{10}=\frac{400}{1,000}$ and $\frac{9}{100}=\frac{90}{1,000}$.

So, we have:

$$0.493 = \frac{4}{10} + \frac{9}{100} + \frac{3}{1,000}$$

$$= \frac{400}{1,000}+\frac{90}{1,000}+\frac{3}{1,000}$$

$$= \frac{493}{1,000}.$$

PRACTICE | Fill in the missing numerators in each equation below.
Then, write each decimal as a fraction.

29. $0.21 = \dfrac{}{10}+\dfrac{}{100} = \dfrac{20}{100}+\dfrac{}{100} = \dfrac{}{100}$

30. $0.74 = \dfrac{}{10}+\dfrac{}{100} = \dfrac{70}{100}+\dfrac{}{100} = \dfrac{}{100}$

31. $0.306 = \dfrac{}{10}+\dfrac{}{100}+\dfrac{}{1,000} = \dfrac{}{1,000}+\dfrac{0}{1,000}+\dfrac{}{1,000} = \dfrac{}{1,000}$

32. $0.041 = \dfrac{}{100}+\dfrac{1}{1,000} = \dfrac{}{1,000}+\dfrac{}{1,000} = \dfrac{}{1,000}$

33. ★ $0.6050 = \dfrac{}{1,000}$

EXAMPLE | Write 0.513 as a fraction.

The method described on the previous page allows us to write

$$0.513 = \frac{5}{10} + \frac{1}{100} + \frac{3}{1,000}$$

$$= \frac{500}{1,000} + \frac{10}{1,000} + \frac{3}{1,000}$$

$$= \frac{513}{1,000}$$

Using this method, we see that we can write any decimal with three digits to the right of the decimal point as a number of thousandths.

> A 1-digit number to the right of a decimal point represents a number of tenths.

> A 2-digit number to the right of a decimal point represents a number of hundredths.

> A 3-digit number to the right of a decimal point represents a number of thousandths, and so on.

$$0.7 = \frac{7}{10}. \qquad 0.71 = \frac{71}{100}. \qquad 0.719 = \frac{719}{1,000}.$$

PRACTICE | Write each decimal below as a fraction with a denominator that is a power of 10.

34. 0.3 = _____

35. 0.11 = _____

36. 0.029 = _____

37. 0.61 = _____

38. 0.444 = _____

39. 0.207 = _____

PRACTICE | Write each fraction below as a decimal.

40. $\frac{89}{100}$ = _____

41. $\frac{89}{1,000}$ = _____

42. $\frac{31}{100}$ = _____

43. $\frac{31}{10,000}$ = _____

EXAMPLE | Write $\frac{289}{100}$ as a decimal.

First, we write $\frac{289}{100}$ as a mixed number: $2\frac{89}{100}$.

Then, since $\frac{89}{100} = 0.89$, we have

$$\frac{289}{100} = 2\frac{89}{100} = \textbf{2.89}.$$

PRACTICE | Write each fraction or mixed number below as a decimal.

44. $\frac{53}{100} =$ _____

45. $4\frac{9}{100} =$ _____

46. $\frac{913}{100} =$ _____

47. $\frac{179}{100} =$ _____

48. $\frac{237}{1,000} =$ _____

49. $\frac{1,013}{1,000} =$ _____

PRACTICE | Write each of the decimals below as mixed numbers.

50. $12.1 =$ _____

51. $12.01 =$ _____

52. $1.121 =$ _____

53. $11.21 =$ _____

54. $2.043 =$ _____

55. $234.3 =$ _____

Try these problems for review!

PRACTICE | Complete the following problems.

56. Write "ninety-six hundredths" as a decimal.

56. _____

57. Write 1.45 as a mixed number *in simplest form*.

57. _____

58. Write 9.075 as a mixed number *in simplest form*.

58. _____

59. Circle the number below that is not equal to the other three.

$$1.087 \qquad \frac{187}{100} \qquad \text{one and eighty-seven hundredths} \qquad \frac{1,870}{1,000}$$

60. ★ Circle the decimal below that is equal to $\frac{3}{8}$.

$$0.38 \qquad 0.35 \qquad 0.375 \qquad 0.385 \qquad 3.8$$

61. ★ Lizzie writes a fraction with numerator 45,078. The denominator of her fraction is a power of ten. She converts her fraction to a decimal. Which of the following could be her result? Circle the best answer.

$$405.78 \qquad 45.780 \qquad 0.4578 \qquad 4.5078 \qquad 4.5780$$

In a **Numberlink** puzzle, the goal is to connect pairs of numbers that are equal.

- Paths may only go up, down, left, or right through squares.
- Paths must begin and end at a number, but they may not pass through squares that contain numbers.
- Only one path may pass through each square.

Below is an example of a Numberlink puzzle and its solution. Each path connects a fraction to its equivalent decimal.

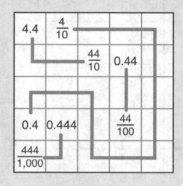

$$\frac{4}{10} = 0.4$$
$$\frac{44}{10} = 4.4$$
$$\frac{44}{100} = 0.44$$
$$\frac{444}{1,000} = 0.444$$

PRACTICE | Solve each Numberlink puzzle below.

62.

	0.7	$7\frac{7}{100}$	$\frac{7}{100}$	$7\frac{77}{100}$
	7.07			
$\frac{7}{10}$			7.77	
0.07		0.77		$\frac{77}{100}$

63.

$\frac{10}{100}$			$\frac{1}{100}$	
	$\frac{11}{100}$		1.01	
0.11				
0.1		$\frac{101}{100}$	0.01	

PRACTICE | Solve each Numberlink puzzle below.

64.

	$\frac{99}{100}$				
		9.99			
9.9					
0.99		$\frac{999}{10}$	$9\frac{99}{100}$	99.9	$\frac{99}{10}$

65.

0.17			$\frac{117}{10}$		
$\frac{107}{100}$	1.7		1.17		
			$\frac{17}{100}$	$1\frac{17}{100}$	11.7
				1.07	
$\frac{17}{10}$					

66.

177.6	14.92				
				1.492	
	1.776	17.76			
	$17\frac{76}{100}$		$\frac{1,776}{1,000}$		
			$\frac{1,776}{10}$		
$\frac{1,492}{1,000}$					$\frac{1,492}{100}$

67.

		31.4	$\frac{314}{1,000}$		
		3.41	0.314		
			3.14		
	$\frac{314}{100}$				
	$\frac{314}{10}$				$\frac{341}{100}$

PRACTICE | Solve each Numberlink puzzle below.

68.

	$\frac{2}{10}$				
		$\frac{22}{100}$	$\frac{22}{10}$		$\frac{222}{100}$
		2.22			0.22
	0.2				2.2
			0.02		$\frac{2}{100}$

69. ★

	$\frac{51}{10}$			1.5	5.1	
	0.15				$\frac{501}{100}$	
				$\frac{15}{10}$	1.05	
			5.01			
	$\frac{15}{100}$					
	1.15			$\frac{115}{100}$	$\frac{105}{100}$	

70. ★★

	5.05		$5\frac{5}{100}$		$\frac{55}{100}$	
		5.50			$\frac{5}{10}$	
	0.05			$\frac{5}{100}$		
	0.55		0.50			
			$\frac{55}{10}$			

71. ★★

	0.33					
				0.333		
		3.33			$\frac{33}{100}$	
		33.3	$\frac{3,333}{100}$		$\frac{333}{10}$	
		$\frac{333}{1,000}$				
$\frac{333}{100}$					33.33

DECIMALS

EXAMPLE | Which is greater: 0.5 or 0.496?

To make 0.5 and 0.496 easier to compare, we write zeros at the end of 0.5 so that both numbers have digits in the same place values:

$$0.5 = 0.500.$$

Then, we write each of 0.500 and 0.496 as fractions:

$$0.500 = \frac{500}{1,000} \quad \text{and} \quad 0.496 = \frac{496}{1,000}.$$

Since $\frac{500}{1,000} > \frac{496}{1,000}$, we have $0.500 > 0.496$.

So, **0.5** is greater than 0.496.

PRACTICE | Place <, >, or = in each circle below to compare each pair of decimals.

72. 0.4 ◯ 0.789

73. 0.151 ◯ 1.49

74. 1.524 ◯ 1.53

75. 5.914 ◯ 5.91

76. 0.064 ◯ 0.64

77. 7.003 ◯ 7.0003

78. Circle the largest decimal number below.

    0.342   0.243   0.234   0.432   0.342   0.423

79. Circle the smallest decimal number below.

    0.02   0.003   0.3   0.002   0.03   0.2

80. Circle the two decimals below that are equal.

    0.78   0.078   0.708   0.807   0.780   0.87

EXAMPLE | Order the following from greatest to least: 0.29, 0.3, 0.045, 0.92, 0.05.

Comparing Decimals

To compare decimals, we first line up the decimal points as shown on the right. This allows us to compare decimals in the same way that we compare whole numbers: we compare digits in place values from left to right.

0.29
0.3
0.045
0.92
0.05

All five numbers have a 0 in the ones place. So, we compare digits in the tenths place.

0.92 has the most tenths, so 0.92 is the greatest.

0.3 has the second-most tenths, and 0.29 has the third-most tenths.

0.92
0.3
0.29
0.045
0.05

We can compare decimals without converting them to fractions!

The remaining numbers are 0.05 and 0.045. Both have a 0 in the tenths place, so we compare their hundredths digits.

Since 0.05 has more hundredths than 0.045, we know 0.05>0.045.

From greatest to least, we have:

0.920
0.300
0.290
0.050
0.045

$$0.92 > 0.3 > 0.29 > 0.05 > 0.045.$$

This ordering is made more obvious by filling empty place values with 0's as shown to the right above.

PRACTICE | Order each set of decimals below from greatest to least on the given lines.

81.   1.70   _____
          0.17
          7.10   _____
          1.07

          _____

          _____

82.   0.409   _____
          0.0494
          0.044   _____
          0.49

          _____

          _____

83.   1.21, 1.12, 1.021, 1.102

          _____

          _____

          _____

          _____

84.   0.675, 0.657, 0.75, 0.6705

          _____

          _____

          _____

          _____

EXAMPLE

Trace a path in the hexagonal grid below that crosses all the numbers on the grid in order from least to greatest.

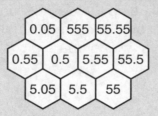

We begin by circling the smallest number on the grid, 0.05. Then, we move from hexagon to hexagon, always connecting to the next-smallest number. We finish at the largest number on the grid, 555.

PRACTICE

Trace a path in each hexagonal grid below that crosses all the numbers on the grid in order from least to greatest.

85.

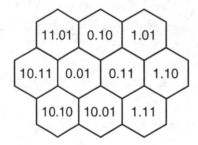

86.

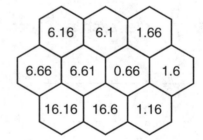

87.

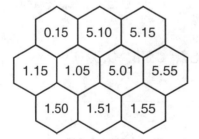

88.

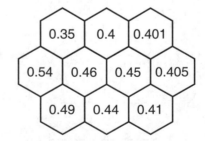

PRACTICE | Trace a path in each hexagonal grid below that crosses all the numbers on the grid in order from least to greatest.

89.

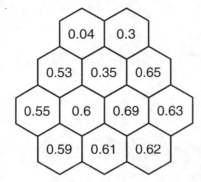

90.

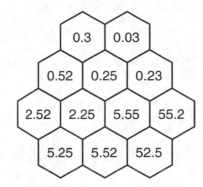

91.

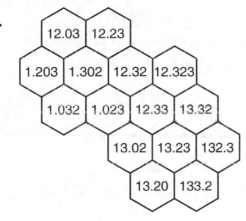

92.

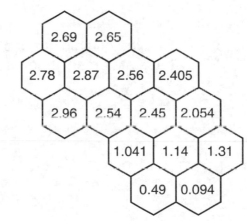

93.

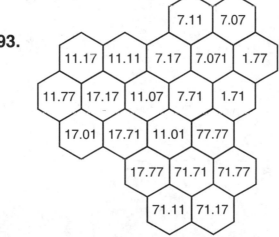

EXAMPLE | Use the digits 5, 6, 7, and 8 once each to fill the blanks below so that the statement is true.

$$7.\boxed{} < \boxed{}.\boxed{} < \boxed{}.7$$

Since the left number is at least 7.0, the ones digits of the middle number $\boxed{}.\boxed{}$ and the right number $\boxed{}.7$ must both be at least 7.

So, the ones digits of the middle and right number must be 7 and 8. Since the number on the right is the largest, we use 8 as its ones digit, and 7 as the ones digit of the middle number:

$$7.\boxed{} < \boxed{7}.\boxed{} < \boxed{8}.7$$

Then, the tenths digit of the left number must be smaller than the tenths digit of the middle number. So, we have

$$\boxed{7}.\boxed{5} < \boxed{7}.\boxed{6} < \boxed{8}.7$$

PRACTICE | Use the given digits once each to fill the blanks in each statement below so that the statement is true.

94. 1, 2, 3, 3

$$\boxed{}.4 < \boxed{}.\boxed{} < \boxed{}.0$$

95. 1, 1, 3, 4

$$\boxed{}.3 < \boxed{}.\boxed{} < 2.1\boxed{}$$

96. 3, 4, 4, 6

$$\boxed{}.5 < \boxed{}.\boxed{} < \boxed{}.2$$

97. 5, 7, 8, 9

$$9.\boxed{}8 < \boxed{}.7 < 9.\boxed{}$$

98. ★ 4, 4, 5, 6, 6, 6

$$5.\boxed{}5 < 5.4\boxed{} < 6 < \boxed{}.6\boxed{} < 6.\boxed{}\boxed{}$$

PRACTICE | For each problem below, use the **same digit** to fill every blank so that the statement or pair of statements is true.

99. 6.☐ < ☐.6 < ☐.☐ < 8.☐

100. ☐.4☐ < ☐.☐ < 6.☐

101. ☐.☐☐ < 5.☐3 < 5.☐☐

102. ☐.5 < 4.☐ & 3.☐ < ☐.4

PRACTICE | For each problem below, each letter represents a **different** digit. Write the digit each letter represents in the blank provided.

103. ★ 5.B8 < A.BC < 6.BC < 6.1C

103. A = _____

B = _____

C = _____

104. ★ K.IJ < 3.IJ < 3.JI < 3.KK

104. I = _____

J = _____

K = _____

105. ★ X.YZ < Z.Y6 < Z.YX < Y.ZX

105. X = _____

Y = _____

Z = _____

EXAMPLE | Write the number marked with an arrow as a decimal.

5.2   5.3

On the number line above, the largest tick marks are labeled to the tenths place (5.2 and 5.3). The arrow points to the third medium tick mark. Since the medium tick marks split one tenth into ten pieces, medium tick marks represent hundredths. We can write 5.2 as 5.20 and 5.3 as 5.30, then label the hundredths between 5.20 and 5.30 as shown below.

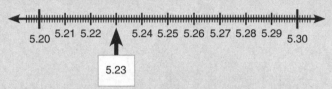

5.20 5.21 5.22   5.24 5.25 5.26 5.27 5.28 5.29 5.30

5.23

So, the marked decimal is **5.23**.

PRACTICE | Label each number marked on the number lines below as a decimal.

106.

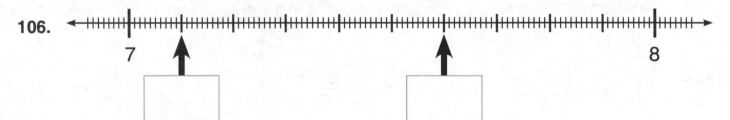

7   8

107.

1.4   1.5

108.

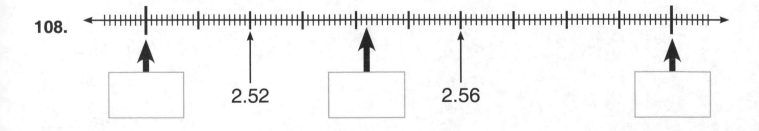

2.52   2.56

PRACTICE | Label each number marked on the number lines below as a decimal.

109.

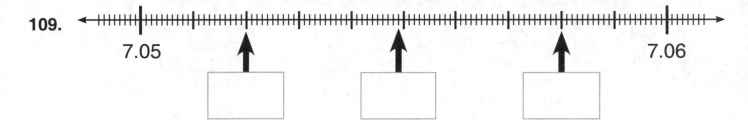

7.05   7.06

110.

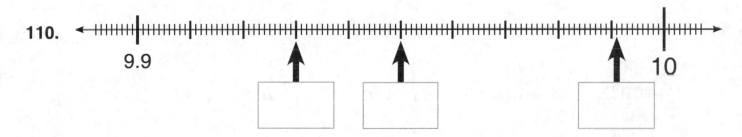

9.9   10

111.

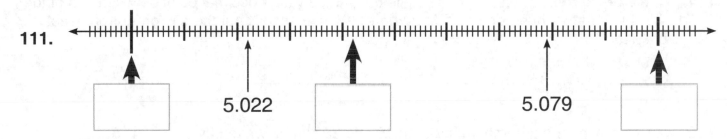

5.022   5.079

PRACTICE | Complete each exercise below.

112. Label each of the numbers 6.39, 6.2, 6.05, 5.8, and 5.63 on the number line below. The first number has been labeled for you.

6   6.39

113. What decimal is the same distance from 6.3 as it is from 6.6 on the
★   number line?

113. _____

EXAMPLE How many different decimal numbers can be created by arranging the digits 3, 4, and 5 and placing a decimal point between two digits?

There are 6 possible arrangements of the digits without the decimal point. From least to greatest, we have:

345	435	534
354	453	543

Then for each arrangement, the decimal point can be placed between the first two digits or between the last two digits. This gives us **12** different decimal numbers.

3.45	4.35	5.34
3.54	4.53	5.43
34.5	43.5	53.4
35.4	45.3	54.3

PRACTICE Answer each question below.

114. There are six ways to arrange the digits 6, 7, and 8 with a decimal point between two of the digits to create a number that is *less than 10*. List these six numbers in order from least to greatest.

_____ < _____ < _____ < _____ < _____ < _____

115. How many different numbers between 10 and 25 can be written by arranging the digits 1, 3, and 5 and inserting a decimal point between two of the digits?

115. _____

116. Arrange the digits 5, 6, and 7 and insert a decimal point between two of the digits to create a number that is between 5.8 and 6.7.

116. _____

117. ★ How many different numbers between 5 and 25 can be written by arranging the digits 2, 3, 4, and 5 and inserting a decimal point between two of the digits?

117. _____

EXAMPLE

Let A, B, and C be digits such that $A>B>C>0$.
Determine whether each of the following are true or false.

$C.B \;\boxed{=}\; C.AB \qquad C.BA \;\boxed{>}\; B.B \qquad B.C \;\boxed{<}\; B.AC$

Since $A>B$, we know that $C.AB$ has more tenths than $C.B$. Since they have the same number of ones, we see that $C.B \;\boxed{<}\; C.AB$. So the first statement is **false**.

Since $C < B$, we know that $C.BA$ has fewer ones than $B.B$. So, $C.BA \;\boxed{<}\; B.B$ and the second statement is also **false**.

Finally, since $A>C$, we know that $B.C$ has fewer tenths than $B.AC$. Since they have the same number of ones, we see that $B.C \;\boxed{<}\; B.AC$. So the third statement is **true**.

PRACTICE

Let A and B be digits such that $A>B>0$.
Fill each circle below with $<$, $>$, or $=$.

118. $0.A \bigcirc 0.B$

119. $0.0A \bigcirc 0.B$

120. $B.B \bigcirc B.AB$

121. $A.0A \bigcirc A.0B$

122. $A.BB \bigcirc B.BA$

123. $0.00B \bigcirc 0.0A$

PRACTICE

Let D, E, and F be digits such that $D>E>F>0$.
Fill each circle below with $<$, $>$, or $=$.

124. $0.DE \bigcirc 0.ED$

125. $0.E \bigcirc 0.0D$

126. $D.EF \bigcirc DE.F$

127. $0.DF \bigcirc 0.DF0$

128. $0.DF \bigcirc 0.EE$

129. $0.0D \bigcirc 0.0E$

DECIMALS

EXAMPLE | Round 3.6275 to the nearest tenth.

3.6275 is between 3.6 and 3.7. Since 3.6275 is closer to 3.6 than to 3.7, we round 3.6275 down to **3.6**.

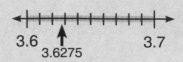

EXAMPLE | Round 3.6275 to the nearest thousandth.

We see that 3.6275 is exactly between 3.627 and 3.628. Numbers that are exactly in the middle are rounded up. So, 3.6275 rounds up to **3.628**.

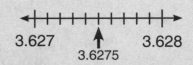

Because numbers in the middle are rounded up, we only need to look at the digit to the right of the place value we are rounding to.

If the digit is a 5, 6, 7, 8, or 9, we round *up*.

If the digit is a 0, 1, 2, 3, or 4, we round *down*.

Include zeros when rounding a number to a place value!

For example, 7.603 rounds to 7.60 when rounded to the nearest hundredth.

PRACTICE | Round each number below to the *nearest tenth*.

130. 1.92 rounds to _____

131. 4.86 rounds to _____

132. 7.01 rounds to _____

133. 13.956 rounds to _____

PRACTICE | Round each number below to the *nearest hundredth*.

134. 9.873 rounds to _____

135. 5.075 rounds to _____

136. 44.444 rounds to _____

137. 49.9999 rounds to _____

Beast Academy Practice 4D

PRACTICE | Round each number below to the **nearest thousandth**.

138. 6.7892 rounds to _____

139. 1.8765 rounds to _____

140. 0.000451 rounds to _____

141. 0.0395 rounds to _____

PRACTICE | Solve each rounding problem below.

142. Fill in the missing digit below so that the resulting number rounds to 0.4 when rounded to the nearest tenth and rounds to 0.45 when rounded to the nearest hundredth.

$$0.4\_5$$

143. Insert a decimal point between two digits of each number below so that each resulting number has tenths digit 4 when rounded to the nearest hundredth.

34653   42396   13947

144. ★ What is the smallest number that rounds to 5.00 when rounded to the nearest hundredth?

144. _____

145. ★ Alex writes every possible arrangement of the digits 3, 4, 5, and 6 with a decimal point between two of the digits. How many of Alex's numbers round to 5 when rounded to the nearest whole number?

145. _____

EXAMPLE | Compute the sum of 9.53 and 6.7.

To add decimals, we stack the numbers, lining up the place values as shown below. Note that we place a zero at the end of 6.7 so that both numbers have digits in the same place values.

$$
\begin{array}{r}
9.53 \\
+\ 6.70 \\
\end{array}
$$

We add the digits in each place value from right to left. Starting with the hundredths, 3+0=3.

$$
\begin{array}{r}
9.53 \\
+\ 6.70 \\
\hline
3 \\
\end{array}
$$

Then, we add the tenths. 5+7=12. We write a 2 in the tenths place and regroup 10 tenths to make 1 one.

$$
\begin{array}{r}
1\ \ \ \\
9.53 \\
+\ 6.70 \\
\hline
23 \\
\end{array}
$$

Finally, we add 1+9+6=16. The decimal point is placed between the ones and the tenths.

$$
\begin{array}{r}
1\ \ \ \\
9.53 \\
+\ 6.70 \\
\hline
16.23 \\
\end{array}
$$

So, 9.53+6.7 = **16.23**.

Check that the decimal points of each number are always lined up!

This will help you make sure your place values match.

PRACTICE | Compute each sum.

146.
$$
\begin{array}{r}
4.31 \\
+\ 5.28 \\
\end{array}
$$

147.
$$
\begin{array}{r}
2.14 \\
+\ 7.37 \\
\end{array}
$$

148.
$$
\begin{array}{r}
9.57 \\
+\ 6.43 \\
\end{array}
$$

149.
$$
\begin{array}{r}
5.381 \\
+\ 0.69 \\
\end{array}
$$

150.
$$
\begin{array}{r}
9.909 \\
+\ 0.101 \\
\end{array}
$$

151.
$$
\begin{array}{r}
4.444 \\
+\ 0.888 \\
\end{array}
$$

Find more practice problems at BeastAcademy.com!

PRACTICE | Compute each sum.

152. $9.21 + 3.07 = \underline{\hspace{2cm}}$

153. $5.84 + 3.19 = \underline{\hspace{2cm}}$

154. $11.2 + 0.43 = \underline{\hspace{2cm}}$

155. $6.75 + 8.056 = \underline{\hspace{2cm}}$

156. $0.3 + 0.55 + 0.777 = \underline{\hspace{2cm}}$

157. $6.6 + 0.77 + 0.088 = \underline{\hspace{2cm}}$

158. Compute $\frac{1}{10} + \frac{23}{100} + \frac{45}{1,000}$. Express your answer as a decimal.

158. \underline{\hspace{2cm}}

159. Compute $0.4 + 0.56 + 0.078$. Express your answer as a mixed number in simplest form.

159. \underline{\hspace{2cm}}

160. Winnie writes every possible arrangement of the digits 3, 5, and 7 with a decimal point between two of the digits. Find the sum of every number less than 5.5 that Winnie writes.

160. \underline{\hspace{2cm}}

EXAMPLE | Subtract 5.83 from 9.6.

To subtract decimals, we stack the numbers, lining up the place values as shown below. Note that we place a zero at the end of 9.6 so that both numbers have digits in the same place values.

$$\begin{array}{r} 9.60 \\ -\ 5.83 \\ \hline \end{array}$$

We cannot take 3 hundredths from 0 hundredths, so we take 1 tenth from the tenths place of 9.60 and break it into 10 hundredths. This gives us 9 ones, 5 tenths, and 10 hundredths to subtract from.

Then, we subtract the hundredths: $10-3=7$. To subtract the tenths, we must take 1 from the ones place and break it into 10 tenths. This gives us 8 ones and 15 tenths to subtract from.

Finally, we subtract the tenths and the ones. $15-8=7$, and $8-5=3$. So, we have $9.6-5.83=\textbf{3.77}$.

$$\begin{array}{r} ^{5\ 10} \\ 9.\cancel{6}\cancel{0} \\ -\ 5.83 \\ \hline \end{array}$$

$$\begin{array}{r} ^{5\ 10} \\ 9.\cancel{6}\cancel{0} \\ -\ 5.83 \\ \hline 7 \end{array}$$

$$\begin{array}{r} ^{8\ 15\ 10} \\ \cancel{9}.\cancel{6}\cancel{0} \\ -\ 5.83 \\ \hline 7 \end{array}$$

$$\begin{array}{r} ^{8\ 15\ 10} \\ \cancel{9}.\cancel{6}\cancel{0} \\ -\ 5.83 \\ \hline \textbf{3.77} \end{array}$$

PRACTICE | Compute each difference.

161.
$$\begin{array}{r} 7.39 \\ -\ 3.26 \\ \hline \end{array}$$

162.
$$\begin{array}{r} 8.34 \\ -\ 2.17 \\ \hline \end{array}$$

163.
$$\begin{array}{r} 9.07 \\ -\ 4.18 \\ \hline \end{array}$$

164.
$$\begin{array}{r} 5.481 \\ -\ 0.86 \\ \hline \end{array}$$

165.
$$\begin{array}{r} 2.005 \\ -\ 0.909 \\ \hline \end{array}$$

166.
$$\begin{array}{r} 5.555 \\ -\ 0.777 \\ \hline \end{array}$$

Find more practice problems at BeastAcademy.com!

PRACTICE | Compute each difference.

167. $9.48 - 3.07 = $ _____

168. $9.84 - 3.29 = $ _____

169. $11.5 - 0.43 = $ _____

170. $7.98 - 4.023 = $ _____

171. $0.5 - 0.009 = $ _____

172. $6.6 - 0.77 = $ _____

173. Subtract $\frac{56}{1,000}$ from $\frac{12}{100}$. Express your answer as a decimal.

173. _____

174. Compute $0.67 - 0.089$. Express your answer as a fraction in simplest form.

174. _____

175. Subtract thirty-one and nineteen hundredths from forty-five and twenty-seven thousandths. Express your answer as a decimal.

175. _____

EXAMPLE | Compute $0.51+0.97+0.13+0.49$.

We could add each term from left to right.

Or, we reorder and regroup terms to make the sum easier to compute. Since $51+49=100$, we have $\frac{51}{100}+\frac{49}{100}=1$. So, $0.51+0.49=1.00=1$. Also, since $97+13=110$, we have $0.97+0.13=1.10=1.1$.

$$0.51+0.97+0.13+0.49$$

Regrouping and adding, we have:

$$0.51+0.97+0.13+0.49=(0.51+0.49)+(0.97+0.13)$$
$$=1.0+1.10$$
$$=\mathbf{2.1.}$$

PRACTICE | Compute each of the following sums.

176. $0.143+1.5+0.157=$ _____

177. $0.11+0.22+0.89+0.78=$ _____

178. $0.25+0.71+0.75=$ _____

179. $1.09+0.41+0.5=$ _____

180. What is the perimeter of the triangle below?

180. _____ cm

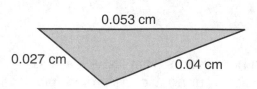

0.053 cm

0.027 cm

0.04 cm

EXAMPLE | Compute 5.14−2.95.

We could line up the digits and subtract.
Or, we could use one of the strategies shown below.

We count up. From 2.95 to 3 is 0.05.
From 3 to 5.14 is 2.14 more.

So, the difference is 0.05+2.14 = **2.19**.

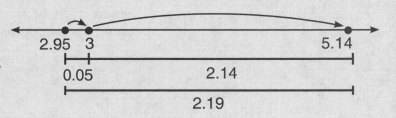

— *or* —

Adding the same value to both
numbers in a subtraction problem
does not change the difference.
We add 0.05 to both 2.95 and 5.14
to make the subtraction easier.

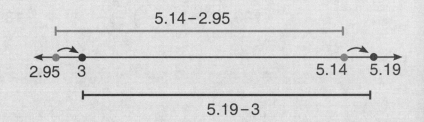

So, 5.14 − 2.95 is equal to (5.14+0.05)−(2.95+0.05) = 5.19−3 = **2.19**.

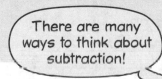
There are many
ways to think about
subtraction!

PRACTICE | Compute each difference below.

181. 6.52−0.98 = _____

182. 4.09−2.88 = _____

183. 1.97−0.99 = _____

184. 3.6−1.84 = _____

185. Compute 2.08+1.65+0.45−0.93.

185. _____

PRACTICE | Place <, >, or = in each circle below to compare each pair of numbers.

186. 1.2 ◯ $\frac{12}{100}$ **187.** $\frac{209}{1,000}$ ◯ 0.290

188. 0.444 ◯ $\frac{55}{100}$ **189.** ★ 0.92 ◯ $\frac{23}{25}$

PRACTICE | Find the perimeter of each rectangle below.

190.
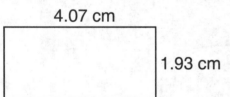
4.07 cm
1.93 cm

190. _____ cm

191.
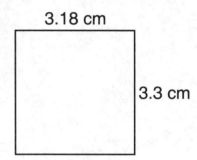
3.18 cm
3.3 cm

191. _____ cm

192.
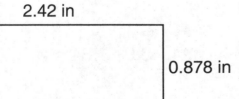
2.42 in
0.878 in

192. _____ in

193.
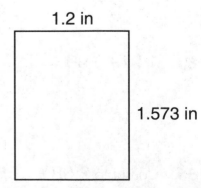
1.2 in
1.573 in

193. _____ in

PRACTICE | Solve each problem below. Remember to include units where necessary.

194. Ray runs the first 17.25 miles of a 26.2-mile marathon in 2 hours and 42 minutes. How much farther must he run to finish the marathon?

194. _____

195. Tok is a monster who weighs 11.42 pounds. Tok would like to weigh 10.5 pounds. How much weight must Tok lose to reach his target weight?

195. _____

196. Dustin cuts a 25 cm string into four pieces. Three of the pieces have lengths 11.59 cm, 4.43 cm, and 4.38 cm. What is the length of the fourth piece of string?

196. _____

197. Circle the decimal below that is closest to 0.17.

$$0.1 \qquad 0.13 \qquad 0.2 \qquad 1.7 \qquad 2.1$$

198. One kilogram weighs about 2.2 pounds. Circle the best estimate
★ of the weight of a 0.75-pound box of cereal in kilograms.

$$0.34 \text{ kg} \qquad 0.43 \text{ kg} \qquad 1.65 \text{ kg} \qquad 2.93 \text{ kg}$$

PRACTICE | Solve each problem below.

199. Shannon writes every number between 0 and 10 that is an arrangement of the digits 4, 5, and 6 with a decimal point between two of the digits.

    a. Find the sum of the smallest and the largest numbers Shannon writes.

        199a. _____

    b. Find the difference between the largest and smallest numbers Shannon writes.

        199b. _____

200. ★ If $A.BC + C.BA = 5.45$, and $A.BC - C.BA = 2.97$, find A, B, and C.

    200. $A = $ _____

           $B = $ _____

           $C = $ _____

201. ★ Fill the missing cells below with decimals so that each row, column, and diagonal has the same sum. The sum of all nine numbers in the grid is 18.9.

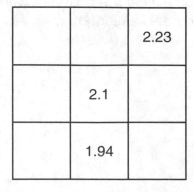

		2.23
	2.1	
	1.94	

PRACTICE Write each decimal below as a fraction *in simplest form*.

202. $0.65 =$ _____

203. $0.675 =$ _____

204. $0.005 =$ _____

205. $0.936 =$ _____

206. ★★ Grogg lists all of the decimals that can be written as $0.AB$ where B is *not zero*. Then, Alex writes all of Grogg's decimals as fractions in simplest form. How many different denominators appear in Alex's list of fractions?

206. _____

207. ★★ Grogg converts $0.ABC$ to a fraction and simplifies the result to $\frac{1}{7}$. Without knowing Grogg's decimal, explain why we know that Grogg must have made an error.

CHAPTER 12
Probability

Use this Practice book with
Guide 4D from BeastAcademy.com.

Recommended Sequence:

Book	Pages:
Guide:	70-83
Practice:	73-81
Guide:	84-109
Practice:	82-101

You may also read the entire chapter
in the Guide before beginning the
Practice chapter.

Probability describes how likely something is to happen.

If an event is *likely* to happen, we say that the probability of the event is *high*. For example, the probability that it will rain somewhere in the United States today is very high.

If an event is *unlikely* to happen, we say that the probability of the event is *low*. For example, the probability that someone you know will win the lottery today is very low.

> Understanding probability helps us make predictions based on what we *expect* to happen.

PRACTICE | Consider the probability of each event below. Draw an arrow from the event on the left to the scale on the right to describe the likelihood of each event.

1. The next stranger you will see has a birthday in April.

2. The next baby born in your town will be a girl.

3. Someone you know will have cereal for breakfast tomorrow.

4. A walrus will bring your favorite uncle a bouquet of roses tomorrow.

5. When you spin a globe, close your eyes, and place your finger somewhere on the globe, your finger will be on a region that is covered by water.

6. You will think about math at least once today.

Certain to happen

Very likely

Likely

Somewhat likely

Equally likely/unlikely

Somewhat unlikely

Unlikely

Very unlikely

Impossible

Many probability problems require counting. Review the Counting chapter in Beast Academy 4B for several important counting techniques that will be used in this chapter.

EXAMPLE | How many whole numbers are between 73 and 115?

The word **between** means that we do not include the first and last numbers in the list. So, we want to know how many numbers are in the list below:

74, 75, 76, ..., 112, 113, 114.

To make the list easier to count, we subtract 73 from each number in the list:

$$
\begin{array}{cccccccc}
74, & 75, & 76, & 77, & ..., & 112, & 113, & 114 \\
-73 & -73 & -73 & -73 & & -73 & -73 & -73 \\
\hline
1, & 2, & 3, & 4, & ..., & 39, & 40, & 41
\end{array}
$$

Now, we have a list of the whole numbers from 1 to 41.

There are 41 numbers in this list.

Since this list is the same size as our original list, we know there are **41** numbers between 73 and 115.

It's easy to tell how many numbers are in a list if we can turn it into a list of whole numbers starting at 1.

PRACTICE | Solve each counting problem below.

7. How many numbers are in the list below?

    13, 14, 15, ... , 234, 235, 236

7. _____

8. How many numbers are in the list below?

    16, 19, 22, ... , 61, 64, 67

8. _____

9. How many even numbers are between 15 and 65?

9. _____

10. How many odd numbers are between 10 and 200?

10. _____

11. How many multiples of 5 are between 101 and 1,001?

11. _____

EXAMPLE Henry, Igor, Jill, and Kurg are racing carts. In how many different ways can the four little monsters finish 1st through 4th in the race?

Any one of the 4 monsters can win. Once a winner is chosen, there are 3 monsters who can place second. Then, one of the 2 remaining monsters will place third. Finally, the 1 remaining monster will place fourth.

So, there are $4 \times 3 \times 2 \times 1 = \textbf{24}$ ways that the four little monsters can place 1st through 4th. They are listed below using the first initial of each monster:

HIJK	IHJK	JHIK	KHIJ
HIKJ	IHKJ	JHKI	KHJI
HJIK	IJHK	JIHK	KIHJ
HJKI	IJKH	JIKH	KIJH
HKIJ	IKHJ	JKHI	KJHI
HKJI	IKJH	JKIH	KJIH

PRACTICE Solve each counting problem below.

12. How many different arrangements of the letters A, B, and C are possible?

12. _____

13. How many arrangements of the digits 1, 2, 3, 4, and 5 create a 5-digit number that is divisible by 5?

13. _____

14. Donna, Ed, Fred, Grace, and Hannah sit in a row of five chairs. If Fred insists on sitting in the middle, how many different arrangements of these five people are possible?

14. _____

15. How many arrangements of the digits 6, 7, 8, and 9 create a 4-digit number that is even?

15. _____

Review counting strategies and tree diagrams in Chapter 4 of Guide 4B.

PRACTICE | Solve each counting problem below. Constructing a tree diagram may be helpful for some problems.

16. Winnie rolls a standard six-sided die, then flips a coin. How many different outcomes are possible? For example, rolling a 4 then flipping tails is one possible outcome.

16. _____

17. Billy is looking at new frames for his glasses. Billy can choose from seven different frame styles, and each style comes in three different colors. How many choices of frames are available?

17. _____

18. At the ice cream shop, there are 9 flavors of ice cream and 13 toppings. How many ways are there to make an ice cream cone with one flavor of ice cream and one topping?

18. _____

19. How many different 4-digit numbers have only odd digits?

19. _____

20. At Tony's Pizzeria, pizzas are available in small, medium, or large. There are 9 topping choices, and each pizza has either thin or thick crust. How many one-topping pizzas are available at Tony's?

20. _____

EXAMPLE | A class has 15 students. How many ways are there to choose two students from the class to clean the blackboard?

You might guess that since there are 15 choices for the first student, and 14 choices for the second student, there are $15 \times 14 = 210$ ways to choose the two students.

However, the order in which we choose the students does not matter! For example, choosing Grogg first and Winnie second creates the same pair as choosing Winnie first and Grogg second. When we multiply 15×14 to get 210, every possible pair of students is counted twice!

Therefore, $15 \times 14 = 210$ is twice the number we want. We must divide by 2 get the correct number of possible pairs.

So, there are $210 \div 2 = \mathbf{105}$ total ways to choose two students to clean the blackboard.

We can count all of the pairs without listing them all.

PRACTICE | Solve each counting problem below.

21. At the smoothie shop, you can select two different fruits to be blended together to make a smoothie. The available fruits are apples, mangos, peaches, strawberries, bananas, grapes, and blueberries. How many two-fruit smoothies are possible? (A banana-grape smoothie is the same as a grape-banana smoothie.)

21. _____

22. An 8-member basketball team must select two team members to be co-captains for the season. How many different pairs of co-captains are possible? (Selecting Alice and Belle is the same as selecting Belle and Alice to act as co-captains.)

22. _____

23. ★ An ice skating team consists of 5 boys and 5 girls. How many ways are there to choose a group of two girls and two boys to perform together?

23. _____

EXAMPLE | Express the probability of rolling a five on a standard six-sided die as a fraction in simplest form.

For an event in which all possible outcomes are equally likely, we can express the probability of a desired event as

$$\frac{\text{Number of Desired Outcomes}}{\text{Number of Possible Outcomes}}.$$

To find the probability of rolling a five, we count 6 equally likely possible rolls (we can roll a ⚀, ⚁, ⚂, ⚃, ⚄, or ⚅). Only 1 of these 6 rolls is a five ⚄. So, the probability of rolling a five on a standard die is $\frac{1}{6}$.

$$\frac{\text{Ways to roll a five}}{\text{Total possible rolls}} = \frac{1}{6}.$$

If we roll a die many times, we expect each face to come up about $\frac{1}{6}$ of the time.

PRACTICE | Answer each question below as a fraction in simplest form.

24. A bag contains 150 jellybeans, 20 of which are black. Katie picks out a jellybean from the bag without looking. What is the probability that Katie's jellybean is black?

24. _____

25. A numbered ball is drawn from a bin containing 75 balls numbered 1 through 75. What is the probability that the number on the ball is odd?

25. _____

26. ★ 🖉 What is the greatest possible probability for any event? What is the least possible probability for any event? Explain.

A standard deck of cards has 52 cards.

There are 4 suits, two of which are black (clubs ♣, spades ♠) and two of which are red (diamonds ◇, hearts ♡).

There are 13 ranks: 2, 3, 4, 5, 6, 7, 8, 9, 10, Jack (J), Queen (Q), King (K), and Ace (A). Each suit has one card of each rank.

In a **shuffled** deck, each card is equally likely to be chosen.

PRACTICE | Answer the following probability questions about a deck of cards. Write your answers as fractions in simplest form.

27. What is the probability that a single card drawn from a shuffled deck of cards is a spade?

27. _____

28. What is the probability of drawing a red card from a shuffled deck of cards?

28. _____

29. What is the probability of drawing a 7 of any suit from a shuffled deck of cards?

29. _____

30. What is the probability of drawing a black Ace from a shuffled deck of cards?

30. _____

31. ★ What is the probability of drawing a card that is red or a King (including a red King)?

31. _____

32. ★ What is the probability of drawing a card that is **neither** a spade nor a 7?

32. _____

If we say that an item is chosen "at random" from a group, we mean that every item in the group has an equal chance of being selected.

EXAMPLE

A two-digit number is selected at random. What is the probability that the number is divisible by 11?

There are 90 two-digit numbers (10 through 99). Nine two-digit numbers are divisible by eleven: 11, 22, 33, ..., 77, 88, 99. So, the probability that a randomly selected two-digit number is divisible by 11 is $\frac{9}{90} = \frac{1}{10}$.

PRACTICE | Answer each question below as a fraction in simplest form.

33. A student is chosen at random from a class of 12 girls and 15 boys. What is the probability that the chosen student is a boy?

33. _____

34. What is the probability that a randomly selected letter of the English alphabet appears in the word PROBABILITY?

34. _____

35. What is the probability that a randomly selected two-digit number has only even digits?

35. _____

36. Grogg randomly picks a positive integer that is less than 20. What is the probability that Grogg's number is prime?

36. _____

37. What is the probability that a word selected randomly from this question has four letters?

37. _____

PRACTICE | Answer each question below as a fraction in simplest form.

38. A number is randomly chosen from the top row shown below. That number is added to a number randomly chosen from the bottom row. What is the probability that the sum is 10?

38. _____

1, 2, 3, 4, 5
5, 6, 7, 8, 9

39. What is the probability that a randomly selected factor of 48 is odd?

39. _____

40. What is the probability that a randomly selected factor of 77 is even?

40. _____

41. A bag contains numbered tiles. The bag contains one tile numbered 1, two tiles numbered 2, three tiles numbered 3, and so on up to ten tiles numbered 10. What is the probability that a randomly selected tile from the bag has a number that is *greater than* 7?

41. _____

42. ★ Olivia creates a list of all the fractions less than 1 that she can write using only the digits 1, 2, 3, 4, and 5, with one digit in the numerator and one digit in the denominator. Dierdra selects one of Olivia's fractions at random. What is the probability that the fraction Dierdra selects is *greater than* $\frac{1}{2}$?

42. _____

A flipped coin is equally likely to land heads or tails. There are 2 equally likely outcomes, and 1 outcome is tails. So, the probability of flipping tails on a fair coin is $\frac{1}{2}$. Similarly, the probability of flipping heads on a fair coin is $\frac{1}{2}$.

What happens when you flip two or more coins? Let's experiment.

Experiment 1: Two coins. Flip both coins at the same time. Place a tally mark in the appropriate column in a chart like the one below for each flip. Do this 60 times.

Two
Heads   Two
Tails   One
Each

Consider: Did all three outcomes occur about the same number of times? If not, which outcome was most likely? Which outcome was least likely? Do you think that your results would be similar if you flipped both coins 60 more times?

Experiment 2: Three coins. Flip three coins at the same time. Place a tally mark in the appropriate column in the chart below for each flip. Do this 60 times.

Two
Heads
&
Three   One   Two Tails & One Heads   Three
Heads   Tails   Tails

Consider: Did all four outcomes occur about the same number of times? If not, which outcome was most likely? Which outcome was least likely? Do you think that your results would be similar if you flipped all three coins 60 more times?

PRACTICE | Use the diagram below to help you answer the questions that follow.

Winnie flips a penny and Grogg flips a nickel. The tree diagram on the right displays all of the possible outcomes for Winnie's penny and Grogg's nickel. (H = Heads, T = Tails)

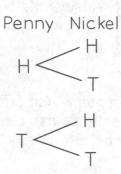

43. How many possible outcomes are possible for the flip of Winnie's penny and Grogg's nickel?

43. _____

44. What is the probability that Winnie's penny will land heads and Grogg's nickel will land tails?

44. _____

45. What is the probability that both Grogg and Winnie will flip heads?

45. _____

46. What is the probability that Grogg's coin will land with the same face up as Winnie's coin?

46. _____

PRACTICE | Answer each question below.

47. Grogg flips two identical coins. What is the probability both coins will land tails?

47. _____

48. Grogg flips his pair of identical coins a total of 1,000 times. About how many of those times do you expect **both** coins to land heads? Circle one answer.

0-50   51-100   101-200   201-300   301-500   501-1,000

PROBABILITY

Coin Flips

PRACTICE | Answer the questions below. Express your answers as fractions in simplest form.

49. Winnie flips a penny, Grogg flips a nickel, and Alex flips a dime. Create a tree diagram to display all of the possible outcomes for Winnie's penny, Grogg's nickel, and Alex's dime.

50. How many possible outcomes are in your tree diagram above?

50. _____

51. What is the probability that all three coins will land heads?

51. _____

52. What is the probability that Winnie and Alex will flip heads, and Grogg will flip tails?

52. _____

53. What is the probability that exactly one little monster will flip heads?

53. _____

PRACTICE | Answer the questions below. Express your answers as fractions in simplest form.

54. Four coins are flipped. What is the probability that all four coins will land heads?

54. _____

55. Four coins are flipped. What is the probability that exactly one coin will land heads?

55. _____

56. Lizzie flips the same coin four times in a row. What is the probability that the third and fourth flips will both land heads?

56. _____

57. ★ Grogg and Lizzie each flip two coins. What is the probability that they each flip the same number of heads?

57. _____

58. Eve flips three identical coins 1,000 times. About how many of those times do you expect **exactly two** coins to land heads?

0-50   51-100   101-200   201-300   301-500   501-1,000

A standard six-sided die is equally likely to land on any of its sides. Since there are 6 equally likely outcomes on a standard die, and 1 of those outcomes is a 4 (⚃), the probability of rolling a 4 on a fair die is $\frac{1}{6}$.

Similarly, the probability of rolling any of the other numbers is $\frac{1}{6}$.

What happens when you roll two or more dice? Let's experiment.

Two dice: With a pair of dice, you can roll any sum from 2 (⚀⚀) to 12 (⚅⚅). Roll a pair of dice 60 times and compute the sum of the faces you have rolled. For each roll, place a tally mark in the appropriate sum column in the chart below.

<u>2</u>   <u>3</u>   <u>4</u>   <u>5</u>   <u>6</u>   <u>7</u>   <u>8</u>   <u>9</u>   <u>10</u>   <u>11</u>   <u>12</u>

Consider: Did all eleven sums occur about the same number of times? If not, which sum was most likely? Which sum was least likely? Do you think that your results would be similar if you rolled the dice 60 more times?

PRACTICE | Alex rolls one die and Grogg rolls a different die. Fill in the entries in the chart below with the possible sums of Alex's and Grogg's rolls. One entry has been completed for you.

59.

Grogg's Die

	1	2	3	4	5	6
1						
2						
3		5				
4						
5						
6						

Alex's Die

PRACTICE | Use the chart you completed on the previous page to answer the questions below.

60. How many outcomes are possible for Grogg's and Alex's rolls? For example, Alex rolling a 3 and Grogg rolling a 4 is one possible outcome. A different possible outcome is Alex rolling a 4 and Grogg rolling a 3.

60. _____

61. In how many different ways can Alex and Grogg roll a sum of 4?

61. _____

62. What is the probability that Alex and Grogg will roll a sum of 4?

62. _____

63. What is the probability that Alex and Grogg will roll a sum of 8?

63. _____

64. **a.** What sum is the most likely?

64. a._____

b. What is the probability of rolling this sum?

b._____

65. What is the probability that the sum of Alex's roll and Grogg's roll will be *less than* 5?

65. _____

66. Pearl rolls a pair of dice 1,200 times. About how many of those times do you expect Pearl to roll a sum of 10?

0-50   51-200   201-300   301-500   501-800   801-1,200

EXAMPLE A point is selected at random within the large triangle below. What is the probability that the selected point is within the shaded area?

The triangle above has 4 congruent regions (3 gray and 1 white). Since 3 of these 4 congruent regions are shaded, $\frac{3}{4}$ of the triangle's area is shaded.

The probability that a randomly selected point in the large triangle is within the shaded area is $\frac{3}{4}$.

PRACTICE A point is selected at random within each outlined shape below. Determine the probability that the randomly selected point is **within a shaded region**.

67. _____

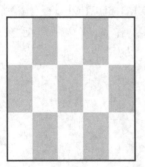

68. _____

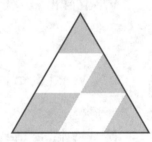

69. _____

70. _____

PRACTICE | Answer each question below.

71. A point is selected at random within the rectangle below. What is the probability that the randomly selected point is within the shaded area?

71. _____

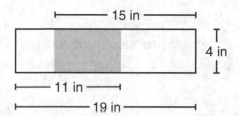

72. Paul throws a dart at a square dartboard with side length 5 cm. The bullseye is a square of side length 2 cm. If Paul's dart strikes a random point on the dartboard, what is the probability that the dart lands in the bullseye?

72. _____

73. Deanna draws a square of side length 1 cm inside a larger square of side length 7 cm. Phil chooses a random point inside the larger square. What is the probability that Phil's point is *not* in Deanna's square?

73. _____

74. ★ ★ ★ Two overlapping rectangles are combined to create the figure below, and the overlapping region is shaded. A point is randomly selected inside the figure. What is the probability that the point is within the shaded area?

74. _____

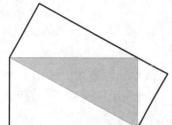

Events are considered *independent* if they do not affect each other. For example, if you flip two coins, the way the first coin lands does not affect the second coin.

EXAMPLE | On five consecutive rolls of a fair, six-sided die, Grogg rolls a 1, 1, 1, 1, and 1. What is the probability that his next roll will be a 1?

The probability of rolling a 1 on six consecutive rolls of a die is very small. So, you might guess that it is very unlikely for Grogg to roll a 1 after rolling a 1 five times.

However, the probability of rolling a 1 on his sixth roll is exactly the same as it was on his first roll: $\frac{1}{6}$.

Grogg's next roll is not affected by his previous rolls. Each roll of the die is an *independent event*.

PRACTICE | Answer each question below.

75. A bag contains 20 discs, numbered 1 through 20. Alex reaches in without looking and draws disc 20. He puts the disc back in the bag, shakes it up, and draws another disc. What is the probability that Alex will draw disc 20 again?

75. _____

76. Inid selects a card from a deck at random, and her selected card is a heart (♡). She replaces the card and shuffles the deck thoroughly. What is the probability that she will draw a heart again at random?

76. _____

77. Lizzie rolls a sum of 2 (⚀⚀) with a pair of standard dice. What is the probability that Lizzie will roll a sum of 2 on her next roll?

77. _____

78. Cammie flips 4 heads in a row with a fair coin. What is the probability that her next three flips will all land heads, for a total of 7 heads in a row?

78. _____

EXAMPLE | The top two cards from a shuffled full deck are turned over, one at a time. The top card is a Jack. What is the probability that the second card is also a Jack?

Two events are **dependent** if the outcome of the first affects the second.

Since the first card was a Jack, we know that there are only 3 Jacks left in the deck. Also, since one card has already been turned over, there are only 51 unseen cards remaining. Since 3 of the 51 unseen cards are Jacks, the probability that the second card is a Jack is $\frac{3}{51} = \frac{1}{17}$.

In this example, the outcome of the first event affects the possible outcomes of the second event because one card has been removed from the deck. So, these two events are **dependent**. We could also say that the outcome of the second event **depends** on the outcome of the first.

PRACTICE | Answer each question below.

79. A bag contains 12 candies: 6 pink and 6 red. Maddi reaches in, pulls out a pink candy, and eats it. Mason then reaches into the bag without looking and pulls out a candy. What is the probability that Mason's candy is also pink?

79. _____

80. Two students are randomly selected from a class of 10 girls and 15 boys. The first student selected is a girl. What is the probability that the second student selected is also a girl?

80. _____

81. Winnie sees that the bottom card in a shuffled standard deck of cards is the 8 of diamonds. What is the probability that the top card in the deck is also a diamond?

81. _____

Many probability problems require careful counting of outcomes.

EXAMPLE A three-digit number is randomly created using only 0's, 1's, and 2's, with repeated digits allowed. What is the probability that the sum of the digits is prime?

First, we count the total number of outcomes. Since 0 cannot be the first digit of a three-digit number, we have two choices for the first digit, three choices for the second digit, and three choices for the third digit. This gives us $2\times3\times3=18$ three-digit numbers.

100	**101**	**102**	**110**	**111**	112	**120**	121	**122**
200	**201**	202	**210**	211	**212**	220	**221**	222

Next, we count the number of successful outcomes. Above, we have bolded the 11 numbers whose digits sum to a prime number.

So, the desired probability is $\frac{11}{18}$.

PRACTICE Answer each probability question below.

82. The digits 1, 2, and 3 are arranged at random and a decimal point is randomly placed **between** two of the digits. What is the probability that the number created is greater than 3?

82. _____

83. The letters O, P, S, and T are arranged at random. What is the probability that the arrangement forms a word in English? For example, OPTS is a word meaning "chooses."

83. _____

84. ★ Three different digits are picked at random from 1, 2, 3, and 4. The three digits are then randomly arranged to make a three-digit number. What is the probability that the three-digit number is divisible by both 2 and 3?

84. _____

85. ★ Two pairs of twins are randomly seated in a row of 4 chairs. What is the probability that each person is sitting next to his or her twin?

85. _____

Creating a tree diagram can help you organize your counting.

EXAMPLE | Jake and Sammy each hold up 1, 2, 3, 4, or 5 fingers at random. What is the probability that the total number of fingers held up is odd?

Each of the boys has 5 choices for the number of fingers he can hold up. So, there are 5×5 = 25 possible ways for each boy to hold up between 1 and 5 fingers as shown below:

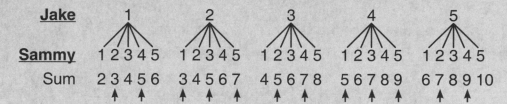

Of these 25 outcomes, the 12 marked with arrows have a total that is odd. So, the probability that the two boys will hold up a total number of fingers that is odd is $\frac{12}{25}$.

PRACTICE | Answer each probability question below. It may help to construct a tree diagram for each problem.

86. Winnie rolls a four-sided die numbered 1 through 4, and Lizzie rolls a standard six-sided die numbered 1 through 6. What is the probability that Lizzie rolls a number that is greater than Winnie's?

86. _____

87. Alicia, Ben, and Corbin each spin one of the spinners below to get a random number from 1 to 3. What is the probability that the product of these three numbers is odd?

87. _____

 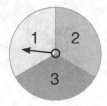 

88. Joe starts with the number one. He changes his number each time he flips a coin. If the result is heads he multiplies his number by three. If the result is tails he multiplies his number by two. What is the probability that after three coin flips, Joe's number is even?

88. _____

If the probability of an event is p, then the probability of that event **not** happening is $1-p$. The **complement** of the event is that event **not** happening.

EXAMPLE

A bag contains blocks that are either blue or red. The probability that a block chosen from the bag is blue is $\frac{5}{7}$. What is the probability that a block chosen from the bag is red?

Since the probability of drawing a blue block is $\frac{5}{7}$, the probability of drawing a block that is **not** blue is $1-\frac{5}{7}=\frac{2}{7}$. Since a block must be red if it is not blue, drawing a red block is the complement of drawing a blue block. The probability of drawing a red block is $\frac{2}{7}$.

> The probability of an event and the probability of its **complement** always add up to 1.

PRACTICE | Answer each question below.

89. Each of the cubes in a bag is labeled with a positive integer. The probability of drawing an odd-numbered cube is $\frac{1}{3}$. What is the probability of drawing an even-numbered cube from the bag?

89. _____

90. For one spin of the spinner shown, the probability that the arrow will land in the 1-point section is $\frac{4}{9}$, and the probability that the spinner will land in the 3-point section is $\frac{2}{9}$. What is the probability that the spinner will land in the 2-point section?

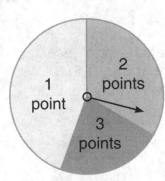

90. _____

91. ★ A bag contains three types of marbles: some red marbles, some black marbles, and some marbles that are red with black stripes. The probability of selecting a marble that has at least some red on it is $\frac{7}{11}$. The probability of selecting a marble that has at least some black on it is $\frac{9}{11}$. What is the probability of selecting a red marble with black stripes?

91. _____

PRACTICE | Answer each question below.

92. **a.** What is the probability of rolling two of the same number with a pair of standard six-sided dice?

92. a._____

b. What is the probability of rolling two different numbers with a pair of standard dice?

b._____

93. **a.** What is the probability of flipping heads on four consecutive flips of a fair coin?

93. a._____

b. What is the probability of flipping **at least one** tails on four consecutive flips of a fair coin?

b._____

94. Rayad picks a number from the list below at random.

100, 101, 102, ... , 298, 299, 300

a. What is the probability that his number is divisible by 7?

94. a._____

b. What is the probability that his number is not divisible by 7?

b._____

A game of chance is **fair** if all players have an equal probability of winning.

EXAMPLE

Tina and Paul each roll one die. If Tina and Paul roll the same number, Tina gets a point. Every time a sum of 7 is rolled, Paul gets a point. The first player to get 10 points wins. Is this game fair?

A game of *chance* is *not fair* if one player is more likely to win than another player.

Looking at the chart for problem 59, we see that there are 6 ways to roll a 7. Since there are 36 total possible rolls, this means that the probability of rolling a 7 is $\frac{6}{36} = \frac{1}{6}$.

Similarly, we see that there are 6 ways to roll doubles. Since there are 36 total possible rolls, this means that the probability of rolling doubles is $\frac{6}{36} = \frac{1}{6}$.

So, the probability of rolling doubles is the same as the probability of rolling a 7. This means that each player has an equal probability of scoring a point, which means each player is equally likely to win. So, **this game is fair.**

PRACTICE | Answer the questions below about games of chance.

95. Amy and Dave each flip a coin. If there is at least one heads, then Dave gets a point. Otherwise, Amy gets a point. The winner of this game is the first person to get 5 points.

a. What is the probability that Dave wins the first point?

95. a._____

b. What is the probability that Amy wins the first point?

b._____

c. Is this game fair? Explain why or why not.

96. Palmer and Richard each roll a standard six-sided die. If the product of their rolls is odd, then Palmer wins. If the product is even, then Richard wins. Is this game fair? Explain why or why not.

PRACTICE | Answer each question below.

97. Larry and James each flip a coin. If both coins land heads, then Larry gets a point. If there is one heads and one tails, then James gets a point. If there are two tails, no points are awarded. The winner is the first person to earn 5 points. Is this game fair? If so, explain why. If not, who has the advantage?

98. Jason flips 1 coin, and Tasha flips 2 coins. If one player flips more heads than the other, the player who flipped more heads gets a point. However, if they flip the same number of heads (including 0), Jason gets a point. The first player to get 10 points wins. Is this game fair? If so, explain why. If not, who has the advantage?

99. Polly spins the spinner on the left, and Molly spins the spinner on the right. If the sum of the two spins is odd, Polly wins. If the sum of the two spins is even, Molly wins. Is this game fair? If so, explain why. If not, who has the advantage?

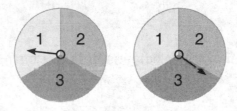

100. Jeremy and Phyllis each roll a standard six-sided die. If they roll different numbers, Phyllis gets one point. If they roll the same number, Jeremy gets *two* points. The first player to get 10 points wins. Is this game fair? If so, explain why. If not, who has the advantage?

EXAMPLE | A bag contains three black blocks, labeled A, B, and C, and two white blocks, labeled X and Y. Two of the blocks are chosen at random. What is the probability that both blocks are black?

Having the blocks labeled with letters helps us organize our thinking. In choosing two blocks, there are 5 choices for the first block and 4 choices for the second block. It would appear that there are $5 \times 4 = 20$ possible pairs of blocks. However, since the pair A&B is the same as the pair B&A, we counted every pair twice. So, there are $20 \div 2 = 10$ possible pairs.

The pairs of black blocks are A&B, A&C, and B&C.
Since 3 of the 10 possible pairs have two black blocks, the probability that both blocks are black is $\frac{3}{10}$.

PRACTICE | Solve each pair-counting probability problem using the information below.

Seven jellybeans are in a bag. Three of the jellybeans are red: cinnamon, cherry, and raspberry. Four of the jellybeans are green: apple, lime, kiwi, and watermelon.

101. How many different pairs of flavors are possible?

101. _____

102. How many different pairs of red jellybeans are possible?

102. _____

103. How many different pairs of green jellybeans are possible?

103. _____

104. Two jellybeans are selected at random from the bag. What is the probability that both jellybeans are red?

104. _____

105. Two jellybeans are selected at random from the bag. What is the probability that both jellybeans are green?

105. _____

106. Two jellybeans are selected at random from the bag. What is the probability that one is red and one is green?

106. _____

EXAMPLE | What is the probability that the top two cards of a shuffled deck are both of the same suit?

There are $52 \times 51 = 2{,}652$ possible ways to pick the top two cards of a deck of cards. We do not want to list all of these possibilities, so we look for a simpler method.

No matter what suit the top card is, 12 of the 51 remaining cards in the deck are the same suit. So, the probability that the second card is the same suit as the first card is $\frac{12}{51} = \frac{4}{17}$.

> Sometimes, a clever insight can help make a tough probability problem much easier.

PRACTICE | All 26 letters of the English alphabet are randomly arranged in a row. Answer the questions below about the arrangement.

107. What is the probability that the 10th letter in the rearranged alphabet is the 10th letter of the standard alphabet (J)?

107. _____

108. What is the probability that the letter A is to the left of the letter Z? (For example, QWERTYUIOP**A**SDFGHJKL**Z**XCVBNM is one such arrangement.)

108. _____

109. ★ What is the probability that the letters X, Y, and Z appear in alphabetical order from left to right? (For example, WS**X**NH**Y**MJUKILOPBGTVFRCDE**Z**AQ is one such arrangement.)

109. _____

110. ★ What is the probability that exactly one of the letters will be in a different position than its position in the standard alphabet?

110. _____

111. ★ Lizzie randomly arranges the letters around a circle instead of a row. What is the probability that M and N are next to each other?

111. _____

The game of **Hats** is played by four people: three students and one teacher.

To begin the game, the teacher places a red or blue hat on the head of each student at random. Each student can see the hats of the other two students, but cannot see his or her own hat. The students may work out a strategy before they are given hats, but *cannot* communicate after they have been given hats.

On the count of three, each student must either guess the color of his or her own hat or stay silent. The teacher wins if everyone stays silent *or* if any student guesses his or her own hat color incorrectly. The students win if at least one student guesses, and if every student who guesses is correct.

For example, if all three students are given blue hats, the teacher wins if everyone is silent or if any student says "red." The students win if at least one of them guesses "blue," and none of them guess "red."

Two examples are shown on the right.

TEACHER WINS	Grogg	Alex	Lizzie
Hat Color	Blue	Blue	Blue
Guess	Silent	"Red"	"Blue"

STUDENTS WIN	Grogg	Alex	Lizzie
Hat Color	Blue	Blue	Blue
Guess	"Blue"	Silent	Silent

What strategy would you use as a student in a game of **Hats**?

Can you find a strategy that allows the students to win more than half of the time?

PRACTICE | Answer the questions below about the game of Hats described above.

112. What is the probability that the students win if one student guesses randomly and the other two stay silent?

112. _____

113. What is the probability that the students win if they all guess randomly?

113. _____

114. ★ What is the probability that the students win if they arrange themselves in a circle and all guess the color of the hat to their right?

114. _____

115. ★ Cammie suggests the following strategy: "Each of us looks at both hats of the other two students. If the two hats you see are the same color, guess that color. Otherwise, stay silent." What is the probability that the students win with Cammie's strategy?

115. _____

116. ★ ✎ Find and describe a way to change Cammie's strategy so that the probability of students winning is *greater than* $\frac{1}{2}$.

PRACTICE | Answer each of the following probability questions.

117. Five cards are chosen at random from a shuffled deck. What is the probability that at least two of them are the same suit?

117. _____

118. What is the probability that the top two cards of a shuffled deck are from different suits?

118. _____

119. ★ James rolls two standard six-sided dice. The sum of the numbers showing on the two dice is 6. What is the probability that both dice show an even number?

119. _____

120. ★ A bag contains 10 marbles: 5 red and 5 black. Ben draws all ten marbles out, one at a time. What is the probability that the tenth marble he pulls out matches the color of the first marble?

120. _____

121. ★ Beginning at point Q, Naymond Newt hops randomly to adjacent points on the given diagram. So, after his first hop, Naymond lands on either point R or point T with equal probability. What is the probability that Naymond lands on point R after his 999th hop?

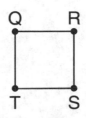

121. _____

HINTS
For Selected Problems

Below are hints to every problem marked with a ★.
Work on the problems for a while before looking at the hints.
The hint numbers match the problem numbers.

CHAPTER 10
Fractions
6

16. After Lisa spends $\frac{1}{6}$ of her money on a book, how much money does she have left?

17. How many minutes does the team spend warming up and doing drills?

24. It may help to start by drawing a diagram. What fraction of the team is made up of 6th and 7th graders?

25. We draw a line to represent Sue's whole race:

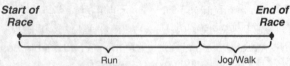

We are given the distance that Sue walked. How far did Sue jog? How far did Sue jog and walk all together?

50. Notice that 54 is a multiple of 9. How could we rearrange the expression to make the product easier to compute?

63. What equation could we write and solve to answer the question "☐ of 20 is 24?"

79. How can we write "$\frac{1}{a}$ of $\frac{1}{b}$ of 100" as a single expression? How can we write "$\frac{1}{b}$ of $\frac{1}{a}$ of 100" as a single expression? Are these two expressions equivalent?

81. How would you locate $\frac{1}{2}$ of $\frac{1}{3}$ on the number line?

84. Use what you learned in Problem 82 to narrow down the possibilities.

85. Use what you learned in Problem 82 to narrow down the possibilities.

86. Given three of the digits, how do we make the first part of the expression, $\left(\frac{\square+\square}{\square}\right)$, as large as possible?

87. Organize your work. Where should the large digits be placed? Where should the small digits be placed?

107. What is the area of the wall Amber is painting?

143. Dividing a by $\frac{1}{16}$ is the same as multiplying a by 16.

149. In the leftmost column, we have $\square \div \frac{1}{36} = 24$. How can we rewrite this equation to make it easier to solve?

163. How can we rewrite the equation $\square \div \frac{1}{5} = 70$ to make it easier to solve?

164. Which ingredient will Laura run out of first?

165. What fraction of Billy's favorite number is $2\frac{1}{2}$?

CHAPTER 11
Decimals
38

33. How could you set this problem up like the previous problems?

60. How can we write each decimal answer choice as a fraction, in simplest form?

61. What would the result be if Lizzie's denominator were 10? How about $10^2 = 100$? $10^3 = 1,000$? What do these results have in common?

69. Begin by pairing each fraction with its equivalent decimal. Then, to make it easier to visualize which numbers must be connected, mark each pair with the same shape or symbol. For example, since $\frac{501}{100} = 5.01$, we mark each of these with a triangle as shown below.

$\frac{51}{10}$			1.5	5.1
	0.15			$\frac{501}{100}$ △
			$\frac{15}{10}$	1.05
		5.01 △		
$\frac{15}{100}$				
	1.15		$\frac{115}{100}$	$\frac{105}{100}$

70. The path that connects 0.55 to its equivalent fraction passes through the center square in the grid.

71. Start with $\frac{333}{100}$. How can you connect $\frac{333}{100}$ to its equivalent decimal, while leaving room to connect $\frac{0,000}{100}$ to its equivalent decimal?

98. What is the ones digit of the fourth number? What is the tenths digit of the first number?

103. Consider $\boxed{6}.\boxed{B}\boxed{C} < \boxed{6}.\boxed{1}\boxed{C}$. What digit is B?

104. Can you order I, J, and K from least to greatest?

105. Can you order X, Y, and Z from least to greatest?

113. How many *hundredths* are between 6.3 and 6.6?

117. The number formed by the digits left of the decimal point is greater than 4 and less than 25. Which digit(s) could be to the left of the decimal point?

144. What is the smallest number that rounds to 5.0 when rounded to the nearest tenth?

145. Fill in the blanks here: Numbers that round to 5 when rounded to the nearest whole number are at least ___, but less than ___.
Which of Alex's numbers are in this range?

189. Can you write each as a number of hundredths?

198. Notice that $0.75+0.75+0.75 = 2.25$, which is pretty close to 2.2.

200. What is the largest possible value of A?

201. The sum of the numbers in each row, column, and diagonal is the same. What is this sum?

206. Try out some possibilities. When we simplify a fraction with denominator 100, what are the possible denominators of the resulting fraction?

207. $0.ABC$ has digits in the tenths, hundredths, and thousandths places, so $0.ABC = \frac{ABC}{1,000}$. Can you write a fraction $\frac{ABC}{1,000}$ that simplifies to $\frac{1}{7}$?

Probability 12

72

23. How many different pairs of girls can be chosen? How many different pairs of boys can be chosen?

26. How could you make $\frac{\text{\# desired outcomes}}{\text{\# possible outcomes}}$ as large as possible? As small as possible?

31. Be sure to count carefully! K♡ and K◇ are both Kings and are both red cards.

32. Be sure to count carefully! One card is both a spade and a 7.

42. Organize the fractions that Olivia writes by their numerators.

57. Create a tree diagram to list all of the possible outcomes.

74. What fraction of each rectangle is shaded?

84. Which groups of three digits could give us a number that is divisible by 3? How many of the arrangements of those groups give us a number that is divisible by 2?

85. Call the first pair of twins A and B, and the second pair of twins X and Y. In how many ways can we arrange A, B, X, and Y? How many of these arrangements have A next to B *and* X next to Y?

91. What is the probability of selecting an all-black marble? An all-red marble?

100. On each roll, how many times more likely is Phyllis to get a point than Jeremy is to get two points?

109. Consider any arrangement of the letters where X, Y, and Z have been removed:

WS☐NH☐MJUKILOPBGTVFRCDE☐AQ.

In how many different ways can the missing letters X, Y, and Z be placed in the empty boxes? How many of these arrangements have X, Y, and Z in alphabetical order?

110. What is the probability that exactly 25 of the letters will be in their standard position?

111. The M is shown on the circle below, with all of the other letters hidden. What is the probability that the N is next to the M?

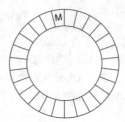

114. There are $2\times2\times2 = 8$ ways to assign the hats to the three students. Consider what each student says in each of the 8 arrangements.

115. There are $2\times2\times2 = 8$ ways to assign the hats to the three students. Consider what each student says in each of the 8 arrangements.

116. Review your work from the previous problem. How could we modify the strategy so that the student losses become wins?

119. What are the possible ways that James could roll a 6 with two dice?

120. What is the probability that the second marble Ben pulls out matches the color of the first?

121. What is the probability that Naymond lands on point R after his first hop? His second hop? His third?

SOLUTIONS
Chapters 10-12

FRACTIONS

"Of"

pages 7-9

1. We split the number line between 0 and 15 into 5 equal pieces. The length of each piece is $\frac{1}{5}$ of 15. The first piece begins at 0 and ends at $\frac{1}{5}$ of 15.

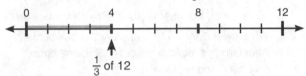

Therefore, $\frac{1}{5}$ of 15 is **3**.

2. We split the number line between 0 and 12 into 3 equal pieces. The length of each piece is $\frac{1}{3}$ of 12.

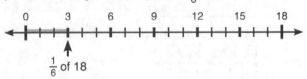

Therefore, $\frac{1}{3}$ of 12 is **4**.

3. We split the number line between 0 and 18 into 6 equal pieces. The length of each piece is $\frac{1}{6}$ of 18.

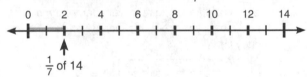

Therefore, $\frac{1}{6}$ of 18 is **3**.

4. We split the number line between 0 and 14 into 7 equal pieces. The length of each piece is $\frac{1}{7}$ of 14.

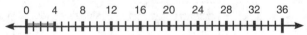

Therefore, $\frac{1}{7}$ of 14 is **2**.

5. We know that one ninth of 36 is 4.

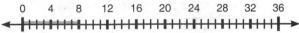

Two ninths of 36 is *two* times as much as one ninth of 36.

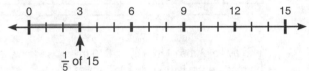

Therefore, $\frac{2}{9}$ of 36 is $2\times4=$ **8**.

We have $\frac{2}{9}\times36 = \left(2\times\frac{1}{9}\right)\times36$
$= 2\times\left(\frac{1}{9}\times36\right)$
$= 2\times4$
$= \textbf{8}.$

6. One third of 36 is 12. So, two thirds of 36 is $2\times12=$ **24**.

$$\frac{2}{3}\times36 = \left(2\times\frac{1}{3}\right)\times36$$
$$= 2\times\left(\frac{1}{3}\times36\right)$$
$$= 2\times12$$
$$= \textbf{24}.$$

7. One fifth of 15 is 3. So, three fifths of 15 is $3\times3=$ **9**.

$$\frac{3}{5}\times15 = \left(3\times\frac{1}{5}\right)\times15$$
$$= 3\times\left(\frac{1}{5}\times15\right)$$
$$= 3\times3$$
$$= \textbf{9}.$$

8. One fourth of 44 is 11. So, three fourths of 44 is $3\times11=$ **33**.

$$\frac{3}{4}\times44 = \left(3\times\frac{1}{4}\right)\times44$$
$$= 3\times\left(\frac{1}{4}\times44\right)$$
$$= 3\times11$$
$$= \textbf{33}.$$

9. One ninth of 18 is 2. So, seven ninths of 18 is $7\times2=$ **14**.

$$\frac{7}{9}\times18 = \left(7\times\frac{1}{9}\right)\times18$$
$$= 7\times\left(\frac{1}{9}\times18\right)$$
$$= 7\times2$$
$$= \textbf{14}.$$

10. One sixth of 18 is 3. So, five sixths of 18 is $5\times3=$ **15**.

$$\frac{5}{6}\times18 = \left(5\times\frac{1}{6}\right)\times18$$
$$= 5\times\left(\frac{1}{6}\times18\right)$$
$$= 5\times3$$
$$= \textbf{15}.$$

11. One seventeenth of 17 is 1. So, five seventeenths of 17 is $5\times1=$ **5**.

$$\frac{5}{17}\times17 = \left(5\times\frac{1}{17}\right)\times17$$
$$= 5\times\left(\frac{1}{17}\times17\right)$$
$$= 5\times1$$
$$= \textbf{5}.$$

12. It is difficult to break 30 into 27 pieces, so we begin by simplifying: $\frac{18}{27}=\frac{2}{3}$. Then, we have:

$$\frac{18}{27}\times30 = \frac{2}{3}\times30$$
$$= \left(2\times\frac{1}{3}\right)\times30$$
$$= 2\times\left(\frac{1}{3}\times30\right)$$
$$= 2\times10$$
$$= \textbf{20}.$$

So, $\frac{18}{27}$ of 30 is **20**.

13. When Peter used $\frac{3}{4}$ of his 16-gallon tank, he used $\frac{3}{4}\times16 = 3\times\left(\frac{1}{4}\times16\right)=3\times4=$ **12** gallons of gas.

14. Two thirds of 12 inches is $\frac{2}{3} \times 12 = $ **8** inches.

15. When Erica used $\frac{2}{3}$ of her 6 gallons of paint, she used $\frac{2}{3} \times 6 = 4$ gallons of paint. So, she has $6 - 4 = $ **2** gallons of paint left.

— *or* —

Erica used $\frac{2}{3}$ of her paint, so she has $\frac{1}{3}$ of her paint left. $\frac{1}{3}$ of 6 gallons is $\frac{1}{3} \times 6 = $ **2** gallons.

16. When Lisa spent $\frac{1}{6}$ of her 36 dollars on a book, she spent $\frac{1}{6} \times 36 = 6$ dollars.

After buying the book, she is left with $36 - 6 = 30$ dollars. She spent $\frac{1}{5}$ of her remaining money on a pencil set, so she spent $\frac{1}{5} \times 30 = 6$ dollars on the pencil set.

All together, Lisa spent $6 + 6 = $ **12** dollars.

17. $\frac{1}{6}$ of 90 minutes is $\frac{1}{6} \times 90 = 15$ minutes.
$\frac{1}{5}$ of 90 minutes is $\frac{1}{5} \times 90 = 18$ minutes.
$\frac{1}{3}$ of 90 minutes is $\frac{1}{3} \times 90 = 30$ minutes.

So, $15 + 18 + 30 = 63$ minutes of practice is spent warming up and doing drills. This leaves $90 - 63 = $ **27** minutes for a scrimmage.

FRACTIONS

Visual Models   10–11

18. We draw a line to represent the total time Ralph spent at the park. Ralph spent 30 minutes playing soccer, which was $\frac{1}{7}$ of the total time. So, the remaining $\frac{6}{7}$ of the time was spent doing other activities.

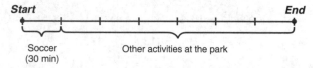

Each seventh of the total time at the park is 30 minutes.

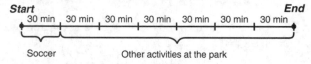

All together, Ralph spent $30 \times 7 = $ **210** minutes at the park this weekend.

19. We draw a diagram to represent Matt's gas tank. When $\frac{1}{5}$ of the tank is full, the remaining $\frac{4}{5}$ of the tank can hold 12 gallons of gas.

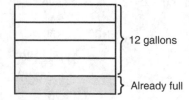

Since *four* fifths of the tank is 12 gallons, we know *each* fifth of the tank is $12 \div 4 = 3$ gallons.

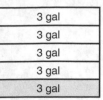

Matt's gas tank holds $3 \times 5 = $ **15** gallons of gas.

20. We draw a line to represent Paul's trip. Since the last $\frac{3}{8}$ of Paul's trip was on a bus, he spent the first $\frac{5}{8}$ of his trip on a train.

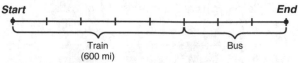

Since *five* eighths of the trip is 600 miles, we know *each* eighth of the trip is $600 \div 5 = 120$ miles.

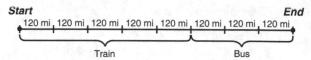

All together, Paul traveled $120 \times 8 = $ **960** miles on his trip.

21. We draw a diagram to represent the box of markers. After Grogg's mom gives out $\frac{1}{4}$ of the markers, $\frac{3}{4}$ of the markers remain.

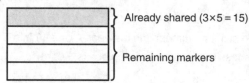

Each fourth of the box contains 15 markers.

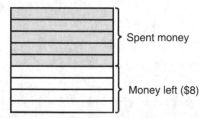

So, the box originally held $4 \times 15 = $ **60** markers.

22. We draw a diagram to represent James's money. Since James spent $\frac{5}{9}$ of his money, he has $\frac{4}{9}$ of his money left.

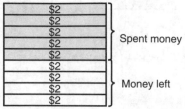

Since *four* ninths of James's money is 8 dollars, we know each ninth of his money is $8 \div 4 = 2$ dollars.

So, *five* ninths of James's money is $5 \times 2 = $ **10** dollars, which he spent on his game.

23. We draw a line to represent the total time Lizzie spent in study hall. Since Lizzie read for $\frac{5}{6}$ of her study hall time, she spent $\frac{1}{6}$ of study hall working on math homework.

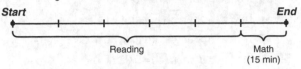

So, each sixth of Lizzie's study hall time is 15 minutes.

All together, she spends $6 \times 15 = \mathbf{90}$ minutes in study hall.

24. We draw a diagram to represent the math team. Since $\frac{2}{7}$ of the team members are 8th graders, the 6th and 7th graders make up $\frac{5}{7}$ of the team.

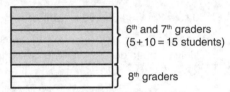

Since *five* sevenths of the team is 15 students, we know *each* seventh of the team is $15 \div 5 = 3$ students.

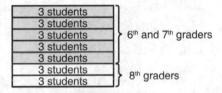

So, *two* sevenths of the team equals $2 \times 3 = \mathbf{6}$ members of the team who are 8th graders.

25. We start by drawing a line to represent Sue's whole race.

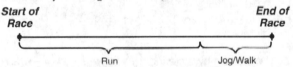

We are given the distance that Sue walked, so we first focus on the part of the race that Sue either jogged or walked.

Since she jogged $\frac{3}{4}$ of the jog/walk distance, she walked for $\frac{1}{4}$ of the jog/walk distance.

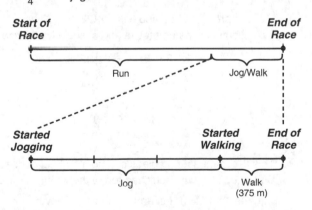

Each fourth of the jog/walk distance is 375 meters.

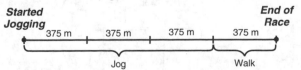

So, the total distance that Sue either walked or jogged is $375 \times 4 = 1,500$ meters.

Now, we look at Sue's whole race. Since Sue ran the first $\frac{7}{10}$ of the race distance, she jogged or walked $\frac{3}{10}$ of the race.

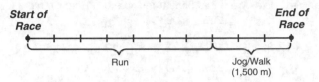

Since three tenths of Sue's race distance is 1,500 meters, we know that each tenth of the race distance is $1,500 \div 3 = 500$ meters.

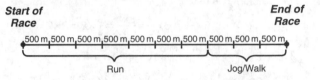

So, Sue's race was $10 \times 500 = \mathbf{5,000}$ meters long.

FRACTIONS

Multiplying Fractions   12-16

26. $5 \times \frac{1}{12} = \frac{1}{12} + \frac{1}{12} + \frac{1}{12} + \frac{1}{12} + \frac{1}{12} = \frac{5}{\mathbf{12}}$.

— *or* —

$5 \times \frac{1}{12} = \frac{5 \times 1}{12} = \frac{5}{\mathbf{12}}$.

27. $8 \times \frac{1}{9} = \frac{8 \times 1}{9} = \frac{\mathbf{8}}{\mathbf{9}}$.

28. $4 \times \frac{2}{3} = \frac{4 \times 2}{3} = \frac{8}{3} = \mathbf{2}\frac{\mathbf{2}}{\mathbf{3}}$.

29. $6 \times \frac{5}{7} = \frac{6 \times 5}{7} = \frac{30}{7} = \mathbf{4}\frac{\mathbf{2}}{\mathbf{7}}$.

30. $8 \times \frac{5}{8} = \frac{8 \times 5}{8} = \frac{40}{8} = \mathbf{5}$.

31. $8 \times \frac{3}{10} = \frac{8 \times 3}{10} = \frac{24}{10} = 2\frac{4}{10} = \mathbf{2}\frac{\mathbf{2}}{\mathbf{5}}$.

— *or* —

$8 \times \frac{3}{10} = \frac{8 \times 3}{10} = \frac{24}{10} = \frac{12}{5} = \mathbf{2}\frac{\mathbf{2}}{\mathbf{5}}$.

32. $2 \times \frac{7}{5} = \frac{2 \times 7}{5} = \frac{14}{5} = \mathbf{2}\frac{\mathbf{4}}{\mathbf{5}}$.

33. $6 \times \frac{10}{7} = \frac{6 \times 10}{7} = \frac{60}{7} = \mathbf{8}\frac{\mathbf{4}}{\mathbf{7}}$.

34. Alex needs $7 \times \frac{2}{3} = \frac{7 \times 2}{3} = \frac{14}{3} = \mathbf{4}\frac{\mathbf{2}}{\mathbf{3}}$ cups of brown sugar to make 7 batches of snickerdoodles.

35. To find $\frac{4}{7}$ of 5, we multiply $\frac{4}{7}$ by 5.

$\frac{4}{7} \times 5 = \frac{4 \times 5}{7} = \frac{20}{7} = \mathbf{2}\frac{\mathbf{6}}{\mathbf{7}}$.

36. $\frac{2}{3}$ of 8 is $\frac{2}{3} \times 8 = \frac{2 \times 8}{3} = \frac{16}{3} = \mathbf{5}\frac{\mathbf{1}}{\mathbf{3}}$.

37. $\frac{5}{8}$ of 3 is $\frac{5}{8} \times 3 = \frac{5 \times 3}{8} = \frac{15}{8} = \mathbf{1}\frac{\mathbf{7}}{\mathbf{8}}$.

38. $\frac{6}{11}$ of 4 is $\frac{6}{11} \times 4 = \frac{6 \times 4}{11} = \frac{24}{11} = \mathbf{2}\frac{\mathbf{2}}{\mathbf{11}}$.

39. $\frac{2}{9} \times 4 = \frac{2 \times 4}{9} = \frac{\mathbf{8}}{\mathbf{9}}$.

40. $\frac{5}{8} \times 7 = \frac{5 \times 7}{8} = \frac{35}{8} = \mathbf{4}\frac{\mathbf{3}}{\mathbf{8}}$.

41. A square field has 4 sides of equal length. So, the perimeter of the field is $\frac{2}{5} \times 4 = \frac{2 \times 4}{5} = \frac{8}{5} = \mathbf{1}\frac{\mathbf{3}}{\mathbf{5}}$ **miles**.

42. $\frac{4}{9}$ of a 7-pound bag of flour weighs $\frac{4}{9} \times 7$ pounds, and seven $\frac{4}{9}$-pound bags of flour weigh $7 \times \frac{4}{9}$ pounds. Multiplication is commutative, so $\frac{4}{9} \times 7 = 7 \times \frac{4}{9}$.

They weigh the same amount!

43. We notice that 55 is a multiple of 11, so we have
$\frac{4}{11} \times 55 = \frac{4 \times 55}{11} = 4 \times \frac{55}{11} = 4 \times 5 = \textbf{20}$.

44. $32 \times \frac{3}{8} = \frac{32 \times 3}{8} = \frac{32}{8} \times 3 = 4 \times 3 = \textbf{12}$.

45. $13 \times \frac{17}{39} = \frac{13 \times 17}{39} = \frac{13}{39} \times 17 = \frac{1}{3} \times 17 = \frac{17}{3} = \textbf{5}\frac{\textbf{2}}{\textbf{3}}$.

46. $99 \times \frac{7}{9} = \frac{99 \times 7}{9} = \frac{99}{9} \times 7 = 11 \times 7 = \textbf{77}$.

47. We first notice that $\frac{10}{15}$ can be simplified to $\frac{2}{3}$. Then we multiply as usual:
$$6 \times \frac{10}{15} = 6 \times \frac{2}{3} = \frac{6 \times 2}{3} = \frac{6}{3} \times 2 = 2 \times 2 = \textbf{4}.$$

48. $12 \times \frac{10}{3} = \frac{12 \times 10}{3} = \frac{12}{3} \times 10 = 4 \times 10 = \textbf{40}$.

49. $\left(66 \times \frac{5}{11}\right) + \left(\frac{4}{13} \times 26\right) = \left(\frac{66}{11} \times 5\right) + \left(4 \times \frac{26}{13}\right)$
$= (6 \times 5) + (4 \times 2)$
$= 30 + 8$
$= \textbf{38}$.

50. We notice that 54 is a multiple of 9, which is the denominator of the first fraction in the expression. Since multiplication is commutative and associative,
$$\frac{8}{9} \times \left(\frac{3}{4} \times 54\right) = \frac{3}{4} \times \left(\frac{8}{9} \times 54\right)$$
$$= \frac{3}{4} \times \left(8 \times \frac{54}{9}\right)$$
$$= \frac{3}{4} \times (8 \times 6).$$

Then, we notice that 8 is a multiple of 4, so we have:
$$\frac{3}{4} \times (8 \times 6) = \left(\frac{3}{4} \times 8\right) \times 6$$
$$= \left(3 \times \frac{8}{4}\right) \times 6$$
$$= (3 \times 2) \times 6$$
$$= \textbf{36}.$$

51. Consider a bag of 44 blocks. If 16 of the 44 blocks are blue, then $\frac{16}{44}$ of the blocks are blue. We simplify $\frac{16}{44}$ to $\frac{4}{11}$.
So, 16 is $\frac{4}{11}$ of 44.
We check that $\boxed{\frac{4}{11}} \times 44 = 16$. ✓

52. If 14 out of 63 blocks are blue, then $\frac{14}{63} = \frac{2}{9}$ of the blocks are blue. So, 14 is $\frac{\textbf{2}}{\textbf{9}}$ of 63.

53. If 31 out of 40 blocks are blue, then $\frac{31}{40}$ of the blocks are blue. So, 31 is $\frac{\textbf{31}}{\textbf{40}}$ of 40.

54. Ted completed $\frac{12}{22} = \frac{\textbf{6}}{\textbf{11}}$ of his homework questions last night.

55. Alice has 27 toy cars and 6 are missing a wheel. So, $27 - 6 = 21$ of her cars are not missing a wheel. Therefore, $\frac{21}{27} = \frac{\textbf{7}}{\textbf{9}}$ of her cars are not missing a wheel.

56. All together, there are $20 + 5 = 25$ students in Nick's class, and 20 have brown eyes. So, $\frac{20}{25} = \frac{\textbf{4}}{\textbf{5}}$ of the students have brown eyes.

57. We use the relationship between multiplication and division: If $\boxed{} \times 8 = 54$, then $\boxed{} = 54 \div 8$.
So, the number we are looking for is $\frac{54}{8} = \frac{27}{4}$, or $\textbf{6}\frac{\textbf{3}}{\textbf{4}}$.
We check that $\boxed{\frac{27}{4}} \times 8 = 54$. ✓

58. If $15 \times \boxed{} = 24$, then $\boxed{} = 24 \div 15$.
So, the number we are looking for is $\frac{24}{15} = \frac{8}{5} = \textbf{1}\frac{\textbf{3}}{\textbf{5}}$.

59. If $21 \times \boxed{} = 9$, then $\boxed{} = 9 \div 21$.
So, the number we are looking for is $\frac{9}{21} = \frac{\textbf{3}}{\textbf{7}}$.

60. If $\boxed{} \times 12 = 20$, then $\boxed{} = 20 \div 12$.
So, the number we are looking for is $\frac{20}{12} = \frac{5}{3} = \textbf{1}\frac{\textbf{2}}{\textbf{3}}$.

61. If $a \times 26 = 30$, then $30 \div 26 = a$.
So, $a = 30 \div 26 = \frac{30}{26} = \frac{15}{13} = \textbf{1}\frac{\textbf{2}}{\textbf{13}}$.

62. If $60 \times b = 35$, then $35 \div 60 = b$. So, $b = 35 \div 60 = \frac{35}{60} = \frac{\textbf{7}}{\textbf{12}}$.

63. To answer "$\boxed{}$ of 20 is 24," we find the number we multiply by 20 to get 24. We write an equation:
$\boxed{} \times 20 = 24$. If $\boxed{} \times 20 = 24$, then $\boxed{} = 24 \div 20$.

So, the number we are looking for is $\frac{24}{20} = \frac{6}{5}$.
24 is $\frac{\textbf{6}}{\textbf{5}}$ of 20.

FRACTIONS
Review 17-20

64. $\frac{3}{5}$ of 16 is $\frac{3}{5} \times 16 = \frac{3 \times 16}{5} = \frac{48}{5}$.
$\frac{2}{5}$ of 10 is $\frac{2}{5} \times 10 = 2 \times \frac{10}{5} = 2 \times 2 = 4$.
So, we want the result of $\frac{48}{5} + 4$.
We write $\frac{48}{5}$ as a mixed number, then add:
$\frac{48}{5} + 4 = 9\frac{3}{5} + 4 = \textbf{13}\frac{\textbf{3}}{\textbf{5}}$.

65. We have $\left(\frac{1}{6} \times 29\right) + \left(\frac{1}{6} \times 31\right) = \frac{29}{6} + \frac{31}{6} = \frac{60}{6} = \textbf{10}$.
— *or* —
We begin by factoring: $\left(\frac{1}{6} \times 29\right) + \left(\frac{1}{6} \times 31\right) = \frac{1}{6} \times (29 + 31)$.
Then, we evaluate: $\frac{1}{6} \times (29 + 31) = \frac{1}{6} \times 60 = \textbf{10}$.

66. The product of 23 and 6 is $23 \times 6 = 138$.
$\frac{5}{3}$ of 138 is $\frac{5}{3} \times 138 = 5 \times \frac{138}{3} = 5 \times 46 = \textbf{230}$.
— *or* —
$\frac{5}{3}$ of the product of 23 and 6 is $\frac{5}{3} \times (23 \times 6)$. Since multiplication is commutative and associative, we have
$$\frac{5}{3} \times (23 \times 6) = \left(\frac{5}{3} \times 6\right) \times 23$$
$$= \left(5 \times \frac{6}{3}\right) \times 23$$
$$= (5 \times 2) \times 23$$
$$= 10 \times 23$$
$$= \textbf{230}.$$

67. This 2-cup measuring glass can hold 16 fluid ounces. When $\frac{5}{8}$ of the cup is full, the top of the water will line up with the $\frac{5}{8} \times 16 = 10$ fluid-once mark, as shown.

68. The total distance between Bobby's and Sammy's houses is 9 cm. When Bobby stops $\frac{7}{10}$ of the way to Sammy's house, he has traveled $\frac{7}{10} \times 9 = \frac{63}{10} = 6\frac{3}{10}$ cm. We mark this distance on the ruler with an arrow as shown:

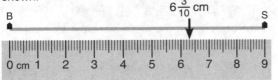

69. Each paperclip is $\frac{5}{8}$ inches long. So, the line of five paperclips is $\frac{5}{8} \times 5 = \frac{25}{8} = 3\frac{1}{8}$ inches long.

We mark this point on the ruler with an arrow as shown:

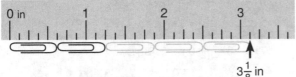

$3\frac{1}{8}$ in

70. Two blocks together weigh $\frac{7}{8}$ of a pound. Eight blocks weigh four times as much as two blocks.

So, eight blocks together weigh $4 \times \frac{7}{8} = \frac{28}{8} = \frac{7}{2} = 3\frac{1}{2}$ pounds. We mark this weight on the second scale as shown:

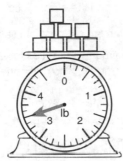

71. From the first scale, we see that the whole pie weighs 4 pounds. So, when $\frac{3}{8}$ of the pie has been eaten, $\frac{3}{8} \times 4 = \frac{12}{8} = \frac{3}{2} = 1\frac{1}{2}$ pounds of pie has been eaten. The weight of the remaining pie is $4 - 1\frac{1}{2} = 2\frac{1}{2}$ pounds.

— *or* —

When $\frac{3}{8}$ of the pie has been eaten, $\frac{5}{8}$ of the pie remains. The whole pie weighs 4 pounds, so the weight of $\frac{5}{8}$ of the pie is $\frac{5}{8} \times 4 = \frac{20}{8} = \frac{5}{2} = 2\frac{1}{2}$ pounds.

We mark this weight on the second scale as shown:

72. $\frac{n}{20} \times 100 = n \times \frac{100}{20} = n \times 5 = 5 \times n$.

73. Winnie gives $\frac{2}{3}$ of her 24 peanuts to Lizzie, so Lizzie gets $\frac{2}{3} \times 24 = 16$ peanuts.

Lizzie gives $\frac{1}{4} \times 16 = 4$ peanuts to Alex.

Alex gives $\frac{1}{2} \times 4 = \mathbf{2}$ peanuts to Grogg.

— *or* —

Grogg received $\frac{1}{2}$ of $\frac{1}{4}$ of $\frac{2}{3}$ of Winnie's 24 peanuts:

$$\frac{1}{2} \times \left(\frac{1}{4} \times \left(\frac{2}{3} \times 24\right)\right) = \frac{1}{2} \times \left(\frac{1}{4} \times 16\right)$$
$$= \frac{1}{2} \times 4$$
$$= 2.$$

So, Grogg gets **2** peanuts.

74. Last Saturday, $\frac{3}{4}$ of the 48 students were in class. So, $\frac{3}{4} \times 48 = 3 \times \frac{48}{4} = 3 \times 12 = 36$ students were in last Saturday's class.

This Saturday, $\frac{1}{6}$ of the 48 students missed class. So, $\frac{1}{6} \times 48 = \frac{48}{6} = 8$ students missed class, and $48 - 8 = 40$ students were in this Saturday's class.

There were $40 - 36 = \mathbf{4}$ more students in class this Saturday than last Saturday.

75. First, $\frac{1}{5}$ of 900 is 180. Then, $\frac{1}{2}$ of 180 is 90.

Finally $\frac{1}{3}$ of 90 is 30. So, $\frac{1}{3}$ of $\frac{1}{2}$ of $\frac{1}{5}$ of 900 is **30**.

— *or* —

We can write an expression:

"$\frac{1}{3}$ of $\frac{1}{2}$ of $\frac{1}{5}$ of 900" means $\frac{1}{3} \times \left(\frac{1}{2} \times \left(\frac{1}{5} \times 900\right)\right)$.

Since multiplication is commutative and associative, we can multiply these in any order. For example,

$$\frac{1}{3} \times \left(\frac{1}{2} \times \left(\frac{1}{5} \times 900\right)\right) = \frac{1}{5} \times \left(\frac{1}{2} \times \left(\frac{1}{3} \times 900\right)\right)$$
$$= \frac{1}{5} \times \left(\frac{1}{2} \times 300\right)$$
$$= \frac{1}{5} \times 150$$
$$= \mathbf{30}.$$

76. We draw the segment of the number line from 0 to n and split it into twelfths. Since *five* twelfths of n is 10, one twelfth of n is $10 \div 5 = 2$.

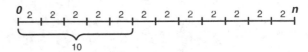

So, seven twelfths of n equals $7 \times 2 = \mathbf{14}$.

77. We draw the segment of the number line from 0 to m and split it into elevenths. If *four* elevenths of m is 12, then *one* eleventh of m is $12 \div 4 = 3$.

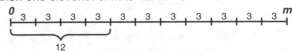

Since one eleventh of m is 3, m equals $11 \times 3 = \mathbf{33}$.

78. $\frac{v}{w} \times w = v \times \frac{w}{w}$. Since w is nonzero, we know $\frac{w}{w} = 1$.

Therefore, $\frac{v}{w} \times w = v \times \frac{w}{w} = v \times 1 = \boldsymbol{v}$.

79. $\frac{1}{a}$ of $\frac{1}{b}$ of 100 is $\frac{1}{a} \times \left(\frac{1}{b} \times 100\right)$.

$\frac{1}{b}$ of $\frac{1}{a}$ of 100 is $\frac{1}{b} \times \left(\frac{1}{a} \times 100\right)$.

Since multiplication is commutative and associative,

$\frac{1}{a} \times \left(\frac{1}{b} \times 100\right) = \frac{1}{b} \times \left(\frac{1}{a} \times 100\right)$.

So, yes, $\frac{1}{a}$ of $\frac{1}{b}$ of 100 is equal to $\frac{1}{b}$ of $\frac{1}{a}$ of 100.

80. $\frac{1}{4}$ of $\frac{1}{5}$ of 100 is $\frac{1}{4} \times \left(\frac{1}{5} \times 100\right) = \frac{1}{4} \times 20 = 5$.

Five is $\frac{5}{100}$ of 100. We simplify $\frac{5}{100} = \frac{1}{20}$, so 5 is $\frac{1}{20}$ of 100.

$\frac{1}{4}$ of $\frac{1}{5}$ of 100 is equal to $\frac{1}{20}$ of 100. Therefore, $n = \boldsymbol{20}$.

81. To compute $\frac{1}{3}$ of y, we split the number line between 0 and y into 3 equal pieces. The length of each piece is $\frac{1}{3}$ of y.

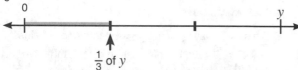

Then, to find $\frac{1}{2}$ of $\left(\frac{1}{3}$ of $y\right)$, we split each of those three pieces into two equal parts. The length of each small part is $\frac{1}{2}$ of $\left(\frac{1}{3}$ of $y\right)$.

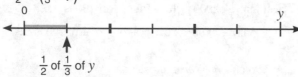

The number line between 0 and y is split into 6 equal parts. So, each part is $\frac{1}{6}$ of y.

No matter what y is, $\frac{1}{2}$ of $\frac{1}{3}$ of y equals $\frac{1}{6}$ of y. Therefore, $z = \boldsymbol{6}$.

FRACTIONS
Digit Arrangements   21

82. To make $\boxed{a} \times \frac{\boxed{b}}{\boxed{c}} = \frac{\boxed{a} \times \boxed{b}}{\boxed{c}}$ as large as possible using the given digits, we place the smallest digit in the denominator and the other digits in the numerator.

The greatest possible value that can be made with the given digits in this arrangement is $\boxed{5} \times \frac{\boxed{3}}{\boxed{2}} = \frac{5 \times 3}{2} = \boldsymbol{7\frac{1}{2}}$.

You may have instead written $\boxed{3} \times \frac{\boxed{5}}{\boxed{2}} = \boldsymbol{7\frac{1}{2}}$.

83. The expression $\frac{\boxed{a}}{\boxed{b}}$ is greater than 1 when $a > b$, but less than 1 when $a < b$.

So, to make $\frac{\boxed{a}}{\boxed{b}} + \boxed{c}$ as large as possible using the given digits, we must have $a > b$.

We test each of the possible digits for c, then place the larger of the two remaining digits in the numerator and the smaller digit in the denominator.

$\frac{7}{6} + 4 = 1\frac{1}{6} + 4 = 5\frac{1}{6}$.

$\frac{7}{4} + 6 = 1\frac{3}{4} + 6 = 7\frac{3}{4}$.

$\frac{6}{4} + 7 = 1\frac{1}{2} + 7 = 8\frac{1}{2}$.

So, $\frac{\boxed{6}}{\boxed{4}} + \boxed{7} = \boldsymbol{8\frac{1}{2}}$ is the greatest possible value.

84. To make $\boxed{a} \times \frac{\boxed{b}}{\boxed{c}} = \frac{\boxed{a} \times \boxed{b}}{\boxed{c}}$ as large as possible using three of the given digits, we place the smallest digit in the denominator and the other digits in the numerator.

So, to find the largest possible value of $\boxed{a} \times \frac{\boxed{b}}{\boxed{c}} + \boxed{d}$, we consider the four possibilities for d, then arrange the remaining three digits with c as small as possible.

$8 \times \frac{7}{4} + 3 = \frac{8}{4} \times 7 + 3 = 2 \times 7 + 3 = 14 + 3 = 17$.

$8 \times \frac{7}{3} + 4 = \frac{56}{3} + 4 = 18\frac{2}{3} + 4 = 22\frac{2}{3}$.

$8 \times \frac{4}{3} + 7 = \frac{32}{3} + 7 = 10\frac{2}{3} + 7 = 17\frac{2}{3}$.

$7 \times \frac{4}{3} + 8 = \frac{28}{3} + 8 = 9\frac{1}{3} + 8 = 17\frac{1}{3}$.

Therefore, $\boxed{8} \times \frac{\boxed{7}}{\boxed{3}} + \boxed{4} = \boxed{7} \times \frac{\boxed{8}}{\boxed{3}} + \boxed{4} = \boldsymbol{22\frac{2}{3}}$ is the greatest possible value.

85. To make $\frac{\boxed{a}}{\boxed{b}} \times \boxed{c} = \frac{\boxed{a} \times \boxed{c}}{\boxed{b}}$ as large as possible using three of the given digits, we place the smallest digit in the denominator and the other digits in the numerator.

So, to find the largest possible value of $\frac{\boxed{a}}{\boxed{b}} \times \boxed{c} - \boxed{d}$, we consider the four possibilities for d, then arrange the remaining three digits to make the largest possible value of $\frac{\boxed{a}}{\boxed{b}} \times \boxed{c}$.

$\frac{7}{3} \times 5 - 2 = \frac{35}{3} - 2 = 11\frac{2}{3} - 2 = 9\frac{2}{3}$.

$\frac{7}{2} \times 5 - 3 = \frac{35}{2} - 3 = 17\frac{1}{2} - 3 = 14\frac{1}{2}$.

$\frac{7}{2} \times 3 - 5 = \frac{21}{2} - 5 = 10\frac{1}{2} - 5 = 5\frac{1}{2}$.

$\frac{5}{2} \times 3 - 7 = \frac{15}{2} - 7 = 7\frac{1}{2} - 7 = \frac{1}{2}$.

Therefore, $\frac{\boxed{7}}{\boxed{2}} \times \boxed{5} - \boxed{3} = \frac{\boxed{5}}{\boxed{2}} \times \boxed{7} - \boxed{3} = \boldsymbol{14\frac{1}{2}}$ is the greatest possible value.

86. To make $\frac{\boxed{a} + \boxed{b}}{\boxed{c}}$ as large as possible using digits a, b, and c, we place the smallest digit in the denominator and the other two digits in the numerator.

So, to find the largest possible value of $\left(\frac{\boxed{a} + \boxed{b}}{\boxed{c}}\right) \times \boxed{d}$, we consider the four possibilities for d, then arrange the remaining three digits to make the largest possible value of $\frac{\boxed{a} + \boxed{b}}{\boxed{c}}$.

$\frac{7 + 6}{5} \times 4 = \frac{13}{5} \times 4 = \frac{52}{5} = 10\frac{2}{5}$.

$\frac{7 + 6}{4} \times 5 = \frac{13}{4} \times 5 = \frac{65}{4} = 16\frac{1}{4}$.

$\frac{7 + 5}{4} \times 6 = \frac{12}{4} \times 6 = 3 \times 6 = 18$.

$\frac{6 + 5}{4} \times 7 = \frac{11}{4} \times 7 = \frac{77}{4} = 19\frac{1}{4}$.

So, $\left(\dfrac{\boxed{6}+\boxed{5}}{\boxed{4}}\right)\times\boxed{7}=\left(\dfrac{\boxed{5}+\boxed{6}}{\boxed{4}}\right)\times\boxed{7}=19\frac{1}{4}$ is the largest possible value.

87. Since we are subtracting, for any digit a the expression $\boxed{a}-\left(\boxed{b}\times\dfrac{\boxed{c}}{\boxed{d}}\right)$ is largest when $\left(\boxed{b}\times\dfrac{\boxed{c}}{\boxed{d}}\right)$ is as small as possible.

To make $\boxed{b}\times\dfrac{\boxed{c}}{\boxed{d}}=\dfrac{\boxed{b}\times\boxed{c}}{\boxed{d}}$ as small as possible using digits b, c, and d, we place the largest digit in the denominator and the other two digits in the numerator.

So, we consider all four possibilities for a, and we arrange the remaining three digits to make the smallest possible value of $\boxed{b}\times\dfrac{\boxed{c}}{\boxed{d}}$.

$3-\dfrac{5\times8}{9}=3-\dfrac{40}{9}$. Since $\dfrac{40}{9}=4\frac{4}{9}$ is greater than 3, we know that $3-\dfrac{40}{9}$ is negative and therefore smaller than any positive result.

$5-\dfrac{3\times8}{9}=5-\dfrac{24}{9}=5-2\frac{2}{3}=2\frac{1}{3}$.

$8-\dfrac{3\times5}{9}=8-\dfrac{15}{9}=8-1\frac{2}{3}=6\frac{1}{3}$.

$9-\dfrac{3\times5}{8}=9-\dfrac{15}{8}=9-1\frac{7}{8}=7\frac{1}{8}$.

So, $\boxed{9}-\left(\boxed{3}\times\dfrac{\boxed{5}}{\boxed{8}}\right)=\boxed{9}-\left(\boxed{5}\times\dfrac{\boxed{3}}{\boxed{8}}\right)=7\frac{1}{8}$ is the largest possible value.

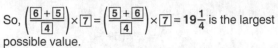

Estimating Products 22-23

88. $\dfrac{31}{210}$ is a little more than $\dfrac{30}{210}$, which simplifies to $\dfrac{1}{7}$.

We circle the answer choice closest to our estimate:

$\dfrac{1}{700}$ $\dfrac{1}{70}$ $\boxed{\dfrac{1}{7}}$ 1

89. $\dfrac{25}{82}$ is a little less than $\dfrac{25}{75}=\dfrac{1}{3}$. So, $\dfrac{25}{82}$ is less than $\dfrac{1}{3}$. Since $\dfrac{2}{3}$, 1, and $\dfrac{5}{3}$ are all greater than $\dfrac{1}{3}$, $\dfrac{25}{82}$ is closest to $\dfrac{1}{3}$.

We circle the answer choice closest to our estimate:

$\boxed{\dfrac{1}{3}}$ $\dfrac{2}{3}$ 1 $\dfrac{5}{3}$

90. $\dfrac{6}{55}$ is a little less than $\dfrac{6}{54}=\dfrac{1}{9}$, and 70 is a little less than 72, which is a multiple of 9. So, we estimate that $\dfrac{6}{55}\times70$ is close to $\dfrac{1}{9}\times72=8$. So, $\dfrac{6}{55}$ of 70 is closest to **8**.

— *or* —

$\dfrac{6}{55}$ is a little more than $\dfrac{6}{60}=\dfrac{1}{10}$. So, we estimate that $\dfrac{6}{55}\times70$ is a little more than $\dfrac{1}{10}\times70=7$.

Among the four answer choices, only **8** fits our estimate. We circle the answer choice closest to our estimate:

2 $\boxed{8}$ 15 22

In fact, $\dfrac{6}{55}\times70=7\frac{7}{11}$.

91. $\dfrac{58}{79}$ is close to $\dfrac{60}{80}=\dfrac{3}{4}$, so we estimate that $\dfrac{58}{79}$ of 24 is about $\dfrac{3}{4}\times24=18$.

We circle the answer choice closest to our estimate:

$2\frac{49}{79}$ $5\frac{49}{79}$ $10\frac{49}{79}$ $\boxed{17\frac{49}{79}}$ $28\frac{49}{79}$

92. $\dfrac{21}{51}$ is close to $\dfrac{20}{50}=\dfrac{2}{5}$, so $\dfrac{21}{51}\times12{,}546\approx\dfrac{2}{5}\times12{,}546$. Since 12,546 is between 10,000 and 15,000, we know $\dfrac{2}{5}\times12{,}546$ is between $\dfrac{2}{5}\times10{,}000=4{,}000$ and $\dfrac{2}{5}\times15{,}000=6{,}000$.

Among the four answer choices, only 5,000 fits our estimate.

$1{,}000$ $\boxed{5{,}000}$ $20{,}000$ $400{,}000$

In fact, $\dfrac{21}{51}\times12{,}546=5{,}166$.

93. $\dfrac{29}{21}$ is between 1 and 2, so $\dfrac{29}{21}\times13{,}982$ is more than $1\times13{,}982=13{,}982$ and less than $2\times14{,}000=28{,}000$. Among the four answer choices, only 20,000 fits our estimate.

$1{,}000$ $5{,}000$ $\boxed{20{,}000}$ $400{,}000$

In fact, $\dfrac{29}{21}\times13{,}982=19{,}308\frac{10}{21}$.

94. $\dfrac{6}{19}$ is a little less than $\dfrac{6}{18}=\dfrac{1}{3}$, and 58 is a little less than 60. So, we estimate that $\dfrac{6}{19}\times58$ is a little less than $\dfrac{1}{3}\times60=20$. Therefore, $\dfrac{6}{19}\times58$ is between **10** and **20**.

In fact, $\dfrac{6}{19}\times58=18\frac{6}{19}$.

95. $\dfrac{25}{109}$ is a little less than $\dfrac{25}{100}=\dfrac{1}{4}$, and 298 is a little less than 300. So, we estimate that $\dfrac{25}{109}\times298$ is a little less than $\dfrac{1}{4}\times300=75$.

Therefore, $\dfrac{25}{109}\times298$ is between **60** and **80**.

In fact, $\dfrac{25}{109}\times298=68\frac{38}{109}$.

96. We estimate each product.

A. $\dfrac{41}{104}\times13$ is about $\dfrac{40}{100}\times10=\dfrac{2}{5}\times13$, which is between $\dfrac{2}{5}\times10=4$ and $\dfrac{2}{5}\times15=6$.

B. $7\times\dfrac{35}{37}$ is a little less than $7\times\dfrac{37}{37}=7\times1=7$.

C. $4\times\dfrac{5}{26}$ is a little less than $4\times\dfrac{5}{25}=4\times\dfrac{1}{5}=\dfrac{4}{5}$.

D. $\dfrac{3}{28}\times33$ is a little more than $\dfrac{3}{30}\times30=3$.

We use these estimates to write the products in order from least to greatest.

$$4\times\dfrac{5}{26}<\dfrac{3}{28}\times33<\dfrac{41}{104}\times13<7\times\dfrac{35}{37}$$

Using the letters instead, we write **C < D < A < B**.

Actual products are given below.

A. $\dfrac{41}{104}\times13=5\frac{1}{8}$ **B.** $7\times\dfrac{35}{37}=6\frac{23}{37}$

C. $4\times\dfrac{5}{26}=\dfrac{10}{13}$ **D.** $\dfrac{3}{28}\times33=3\frac{15}{28}$

Multiplying Mixed Numbers 24-25

97. We convert $1\frac{1}{4}$ to a fraction: $1\frac{1}{4}=\dfrac{5}{4}$.

Then, we multiply: $7\times\dfrac{5}{4}=\dfrac{7\times5}{4}=\dfrac{35}{4}$.

We write $\dfrac{35}{4}$ as a mixed number: $\dfrac{35}{4}=8\frac{3}{4}$.

— *or* —

Fractions Chapter 10 Solutions

We use the distributive property. $1\frac{1}{4}$ equals $1+\frac{1}{4}$, so $7\times1\frac{1}{4}$ equals $7\times\left(1+\frac{1}{4}\right)$. We distribute the 7 and get

$$7\times1\frac{1}{4}=7\times\left(1+\frac{1}{4}\right)$$
$$=\left(7\times1\right)+\left(7\times\frac{1}{4}\right)$$
$$=7+\frac{7}{4}$$
$$=7+1\frac{3}{4}$$
$$=8\frac{3}{4}.$$

98. We use the distributive property to get

$$15\times2\frac{1}{3}=15\times\left(2+\frac{1}{3}\right)$$
$$=\left(15\times2\right)+\left(15\times\frac{1}{3}\right)$$
$$=30+\frac{15}{3}$$
$$=30+5$$
$$=35.$$

99. $5\frac{3}{4}\times9=\left(5+\frac{3}{4}\right)\times9$
$$=\left(5\times9\right)+\left(\frac{3}{4}\times9\right)$$
$$=45+\frac{27}{4}$$
$$=45+6\frac{3}{4}$$
$$=51\frac{3}{4}.$$

100. $8\frac{5}{6}\times11=\left(8+\frac{5}{6}\right)\times11$
$$=\left(8\times11\right)+\left(\frac{5}{6}\times11\right)$$
$$=88+\frac{55}{6}$$
$$=88+9\frac{1}{6}$$
$$=97\frac{1}{6}.$$

101. $9\times6\frac{1}{11}=9\times\left(6+\frac{1}{11}\right)$
$$=\left(9\times6\right)+\left(9\times\frac{1}{11}\right)$$
$$=54+\frac{9}{11}$$
$$=54\frac{9}{11}.$$

102. $3\frac{4}{7}\times12=\left(3+\frac{4}{7}\right)\times12$
$$=\left(3\times12\right)+\left(\frac{4}{7}\times12\right)$$
$$=36+\frac{48}{7}$$
$$=36+6\frac{6}{7}$$
$$=42\frac{6}{7}.$$

103. Wilson's dad's height is

$$33\frac{3}{4}\times2=\left(33+\frac{3}{4}\right)\times2$$
$$=\left(33\times2\right)+\left(\frac{3}{4}\times2\right)$$
$$=66+\frac{6}{4}$$
$$=66+1\frac{1}{2}$$
$$=67\frac{1}{2}\text{ inches.}$$

104. The perimeter of the dodecagon is

$$12\times5\frac{3}{8}=12\times\left(5+\frac{3}{8}\right)$$
$$=\left(12\times5\right)+\left(12\times\frac{3}{8}\right)$$
$$=60+\frac{36}{8}$$
$$=60+\frac{9}{2}$$
$$=60+4\frac{1}{2}$$
$$=64\frac{1}{2}\text{ inches.}$$

105. Each week, Alex runs
$$1\frac{3}{4}\times2=\left(1\times2\right)+\left(\frac{3}{4}\times2\right)=2+\frac{6}{4}=2+1\frac{1}{2}=3\frac{1}{2}\text{ miles.}$$

So, over four weeks, he runs
$$3\frac{1}{2}\times4=\left(3\times4\right)+\left(\frac{1}{2}\times4\right)=12+\frac{8}{4}=12+2=\mathbf{14}\text{ miles.}$$

— *or* —

All together, Alex runs $\left(1\frac{3}{4}\times2\right)\times4$ miles. Since multiplication is associative, we have

$$1\frac{3}{4}\times\left(2\times4\right)=1\frac{3}{4}\times8$$
$$=\left(1\times8\right)+\left(\frac{3}{4}\times8\right)$$
$$=8+\left(3\times\frac{8}{4}\right)$$
$$=8+6$$
$$=\mathbf{14}\text{ miles.}$$

106. Since three bags together weigh 8 pounds, each bag weighs $8\div3=\frac{8}{3}=2\frac{2}{3}$ pounds.

So, ten bags of ice together weigh

$$10\times2\frac{2}{3}=10\times\left(2+\frac{2}{3}\right)$$
$$=\left(10\times2\right)+\left(10\times\frac{2}{3}\right)$$
$$=20+\frac{20}{3}$$
$$=20+6\frac{2}{3}$$
$$=\mathbf{26\frac{2}{3}}\text{ pounds.}$$

— *or* —

We know the weight of three bags. So, to find the weight of ten bags, we consider the weight of 3 groups of three bags, plus the weight of one more bag.

3 groups of three bags weigh $3\times8=24$ pounds, and a single bag weighs $8\div3=\frac{8}{3}=2\frac{2}{3}$ pounds.

All together, the ten bags weigh $24+2\frac{2}{3}=\mathbf{26\frac{2}{3}}$ pounds.

107. The area of Amber's rectangular wall is

$$15\times18\frac{2}{3}=15\times\left(18+\frac{2}{3}\right)$$
$$=\left(15\times18\right)+\left(15\times\frac{2}{3}\right)$$
$$=270+\left(\frac{15}{3}\times2\right)$$
$$=270+\left(5\times2\right)$$
$$=270+10$$
$$=280\text{ square feet.}$$

Amber needs 1 quart of paint for every 75 square feet of wall, so she will need $280\div75=\frac{280}{75}=\frac{56}{15}=3\frac{11}{15}$ quarts to paint 280 square feet of wall.

If you assumed that Amber must use *whole* quarts of paint, you may have rounded up your result to **4** whole quarts to paint the wall.

FRACTIONS
Perimeter and Area   26-27

108. A regular pentagon has five sides of equal length. The perimeter of the given regular pentagon is
$$5 \times 4\frac{3}{8} = \left(5 \times 4\right) + \left(5 \times \frac{3}{8}\right)$$
$$= 20 + \frac{15}{8}$$
$$= 20 + 1\frac{7}{8}$$
$$= \mathbf{21\frac{7}{8}} \textbf{ cm.}$$

109. The area of the rectangle is
$$2\frac{4}{5} \times 6 = \left(2 \times 6\right) + \left(\frac{4}{5} \times 6\right)$$
$$= 12 + \frac{24}{5}$$
$$= 12 + 4\frac{4}{5}$$
$$= \mathbf{16\frac{4}{5}} \textbf{ sq km.}$$

110. The area of each congruent right triangle is *half* the product of leg lengths. First, we multiply the leg lengths:
$$4\frac{2}{3} \times 9 = \left(4 \times 9\right) + \left(\frac{2}{3} \times 9\right)$$
$$= 36 + 6$$
$$= 42.$$
So, the area of each triangle is $42 \div 2 = 21$ square inches. The shaded region is made of four congruent triangles, so the area of the shaded region is $4 \times 21 = \textbf{84 sq in}$.

111. The rectangle and triangle have the same height: $12 \div 2 = 6$ meters. The area of the rectangle is therefore
$$6 \times 3\frac{1}{2} = \left(6 \times 3\right) + \left(6 \times \frac{1}{2}\right)$$
$$= 18 + 3$$
$$= 21 \text{ square meters.}$$
The area of a triangle with the same height and base length as the rectangle is *half* the area of the rectangle: $21 \div 2 = 10\frac{1}{2}$ square meters.
So, the area of the pentagon is $21 + 10\frac{1}{2} = \mathbf{31\frac{1}{2}} \textbf{ sq m.}$

112. The length of the short side of each congruent rectangle is $11 \div 2 = 5\frac{1}{2}$ inches.

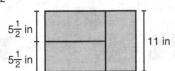

Therefore, the area of each small rectangle is
$$11 \times 5\frac{1}{2} = \left(11 \times 5\right) + \left(11 \times \frac{1}{2}\right)$$
$$= 55 + \frac{11}{2}$$
$$= 55 + 5\frac{1}{2}$$
$$= 60\frac{1}{2} \text{ sq in.}$$

So, the area of the larger rectangle is
$$3 \times 60\frac{1}{2} = \left(3 \times 60\right) + \left(3 \times \frac{1}{2}\right)$$
$$= 180 + \frac{3}{2}$$
$$= 180 + 1\frac{1}{2}$$
$$= \mathbf{181\frac{1}{2}} \textbf{ sq in.}$$

— *or* —

The length of the short side of each small rectangle is $11 \div 2 = 5\frac{1}{2}$ inches, so the length of the long side of the large rectangle is $11 + 5\frac{1}{2} = 16\frac{1}{2}$ inches.

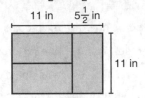

So, the area of the larger rectangle is
$$16\frac{1}{2} \times 11 = \left(16 \times 11\right) + \left(\frac{1}{2} \times 11\right)$$
$$= 176 + \frac{11}{2}$$
$$= 176 + 5\frac{1}{2}$$
$$= \mathbf{181\frac{1}{2}} \textbf{ sq in.}$$

113. Since the perimeter of each triangle is 5 feet, the length of each side of a triangle is $5 \div 3 = \frac{5}{3} = 1\frac{2}{3}$ feet.

The perimeter of the quadrilateral is made up of seven triangle sides, so its perimeter is
$$7 \times 1\frac{2}{3} = \left(7 \times 1\right) + \left(7 \times \frac{2}{3}\right)$$
$$= 7 + \frac{14}{3}$$
$$= 7 + 4\frac{2}{3}$$
$$= \mathbf{11\frac{2}{3}} \textbf{ feet.}$$

— *or* —

Since the perimeter of each triangle is 5 feet, we know each group of three triangle sides is 5 feet long.

The perimeter of the quadrilateral is made up of seven triangle sides. Six triangle sides are $5 \times 2 = 10$ feet long. One triangle side is $5 \div 3 = 1\frac{2}{3}$ feet long.

So, the length of seven triangle sides is
$10 + 1\frac{2}{3} = \mathbf{11\frac{2}{3}} \textbf{ feet.}$

FRACTIONS
Division   28

114. Since each patty weighs $\frac{1}{3}$ of a pound, we know that 3 patties together weigh one pound. So, a 12-pound box of patties contains $12 \times 3 = \textbf{36}$ patties.

115. Since each lap is $\frac{1}{5}$ of a mile, we know that Kayla runs 5 laps to complete one mile. So, to complete 7 miles, Kayla runs $7 \times 5 = \textbf{35}$ laps around the track.

116. Since each sheet of plywood is $\frac{1}{4}$ of an inch thick, we know that a stack of 4 sheets is one inch tall. Therefore, an 18-inch stack of sheets contains $18 \times 4 = \textbf{72}$ sheets.

117. Since each side of the polygon is $\frac{1}{5}$ inches, we know that 5 sides together equal one inch. So, this polygon with perimeter 4 inches must have $4 \times 5 = 20$ sides.

A 20-sided polygon is called an *icosagon*.

118. Since each bag contains $\frac{1}{8}$ of a pound of peppermint bark, we know that 8 bags together make one pound. So, the 50 pounds of bark was split into $8 \times 50 = 400$ bags.

Each bag is sold for $6, so the Candy Shop collects a total of $6 \times 400 = \textbf{2,400}$ dollars.

Reciprocals   29

119. $\frac{1}{5} \times 5 = 1$, so **5** and $\frac{1}{5}$ are reciprocals.

120. $17 \times \frac{1}{17} = 1$, so 17 and $\frac{1}{17}$ are reciprocals.

121. $5 + 6 = 11$, and $11 \times \frac{1}{11} = 1$. So, the reciprocal of $5+6$ is $\frac{1}{11}$.

You may have also noticed that we do not need to evaluate $5+6$ to see that $(5+6) \times \frac{1}{5+6} = 1$. So, we can write the reciprocal of $5+6$ as $\frac{1}{5+6}$.

122. $6 \times 9 = 54$, and $54 \times \frac{1}{54} = 1$. So, the reciprocal of 6×9 is $\frac{1}{54}$.

You may have also noticed that we do not need to evaluate 6×9 to see that $(6 \times 9) \times \frac{1}{6 \times 9} = 1$. So, we can write the reciprocal of 6×9 as $\frac{1}{6 \times 9}$.

123. We first simplify: $\frac{15}{3} = 5$. Then, we have $5 \times \frac{1}{5} = 1$, so $\frac{1}{5}$ is the reciprocal of $\frac{15}{3}$.

You may also notice that we can rewrite $\frac{1}{5}$ as an equivalent fraction, such as $\frac{3}{15}$.

124. We first simplify: $\frac{13}{39} = \frac{1}{3}$. Then, we have $\frac{1}{3} \times 3 = 1$, so **3** is the reciprocal of $\frac{13}{39}$.

You may also notice that we can write 3 as an equivalent fraction, such as $\frac{39}{13}$.

125. The reciprocal of a number n is the number we multiply by n to get 1. **The result of multiplying any number by 0 is 0, so we cannot multiply any number by 0 to get 1. Therefore, 0 has no reciprocal.**

126. We know that $n \times \frac{1}{n} = \frac{n}{n} = 1$ for all nonzero values of n. So, the reciprocal of n is $\frac{1}{n}$.

127. We know that $\frac{1}{a} \times a = \frac{a}{a} = 1$ for all nonzero values of a. So, the reciprocal of $\frac{1}{a}$ is a.

128. We have $(c+1) \times \frac{1}{c+1} = \frac{c+1}{c+1} = 1$. So, the reciprocal of $c+1$ is $\frac{1}{c+1}$.

129. The reciprocal of $\frac{1}{5}$ is 5, and the reciprocal of $\frac{1}{7}$ is 7. Their sum is $5 + 7 = \textbf{12}$.

Division by Unit Fractions   30-31

130. Consider dividing 5 pounds of flour into $\frac{1}{7}$-pound bags. From each pound of flour, we can make seven $\frac{1}{7}$-pound bags. So, from five pounds of flour, we can make $5 \times 7 = \textbf{35}$ bags.

— *or* —

We look at the number line to find out how many $\frac{1}{7}$'s are in 5. Since there are 7 sevenths in 1, there are $5 \times 7 = \textbf{35}$ sevenths in 5.

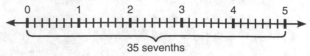

35 sevenths

We write $5 \div \frac{1}{7} = 5 \times 7 = \textbf{35}$.

131. $3 \div \frac{1}{16} = 3 \times 16 = \textbf{48}$.

132. $9 \div \frac{1}{4} = 9 \times 4 = \textbf{36}$.

133. $\frac{1}{16} \div \frac{1}{8} = \frac{1}{16} \times 8 = \frac{8}{16} = \frac{1}{2}$.

134. $3\frac{2}{11} \div \frac{1}{2} = 3\frac{2}{11} \times 2 = 6\frac{4}{11} = \frac{70}{11}$.

135. $2\frac{1}{5} \div \frac{1}{8} = 2\frac{1}{5} \times 8 = (2 \times 8) + \left(\frac{1}{5} \times 8\right) = 16 + \frac{8}{5} = 16 + 1\frac{3}{5}$
$= 17\frac{3}{5} = \frac{88}{5}$.

136. $5 \div \left(3 \div \frac{1}{12}\right) = 5 \div (3 \times 12) = 5 \div 36 = \frac{5}{36}$.

137. $(5 \div 3) \div \frac{1}{12} = \frac{5}{3} \div \frac{1}{12} = \frac{5}{3} \times 12 = \frac{60}{3} = \textbf{20}$.

138. $\left(9 \div \frac{1}{10}\right) \div \frac{1}{5} = (9 \times 10) \div \frac{1}{5} = 90 \div \frac{1}{5} = 90 \times 5 = \textbf{450}$.

139. $9 \div \left(\frac{1}{10} \div \frac{1}{5}\right) = 9 \div \left(\frac{1}{10} \times 5\right) = 9 \div \frac{5}{10} = 9 \div \frac{1}{2} = 9 \times 2 = \textbf{18}$.

140. Each scoop is $\frac{1}{4}$ of a cup of flour, so it will take $2\frac{3}{4} \div \frac{1}{4} = 2\frac{3}{4} \times 4 = (2 \times 4) + \left(\frac{3}{4} \times 4\right) = 8 + 3 = \textbf{11}$ scoops to make $2\frac{3}{4}$ cups.

141. The tree grows $\frac{1}{8}$ of an inch each week, so it will take $7\frac{3}{4} \div \frac{1}{8} = 7\frac{3}{4} \times 8 = (7 \times 8) + \left(\frac{3}{4} \times 8\right) = 56 + 6 = \textbf{62}$ weeks to grow $7\frac{3}{4}$ inches.

142. Tara shared $4 - 1\frac{1}{3} = 2\frac{2}{3}$ gallons of water. Since she shared $\frac{1}{3}$ gallons with each companion, she had $2\frac{2}{3} \div \frac{1}{3} = 2\frac{2}{3} \times 3 = (2 \times 3) + \left(\frac{2}{3} \times 3\right) = 6 + 2 = \textbf{8}$ hiking companions, not including herself.

143. The equation $a \div \frac{1}{16} = 20$ means the same as $a \times 16 = 20$. If $a \times 16 = 20$, then $a = 20 \div 16 = \frac{20}{16} = \frac{5}{4} = 1\frac{1}{4}$.

We check that $\boxed{\frac{5}{4}} \div \frac{1}{16} = \frac{5}{4} \times 16 = 20$. ✓

Cross Number Puzzles   32-33

144. Step 1:

6	×	$\frac{2}{5}$	=	$2\frac{2}{5}$
×		×		×
$1\frac{1}{3}$	×	18	=	24
=		=		=
8	×	$7\frac{1}{5}$	=	

Final:

6	×	$\frac{2}{5}$	=	$2\frac{2}{5}$
×		×		×
$1\frac{1}{3}$	×	18	=	24
=		=		=
8	×	$7\frac{1}{5}$	=	$57\frac{3}{5}$

145. Step 1:

60	×	$\frac{1}{30}$	=	2
÷		÷		×
6	÷		=	30
=		=		=
10	÷		=	60

Final:

60	×	$\frac{1}{30}$	=	2
÷		÷		×
6	÷	$\frac{1}{5}$	=	30
=		=		=
10	÷	$\frac{1}{6}$	=	60

146. Step 1:

3	×	4	=	12
÷		÷		÷
$\frac{1}{35}$	×	7	=	$\frac{1}{5}$
=		=		=
105	×		=	

Final:

3	×	4	=	12
÷		÷		÷
$\frac{1}{35}$	×	7	=	$\frac{1}{5}$
=		=		=
105	×	$\frac{4}{7}$	=	60

147. Step 1:

18	×	$1\frac{2}{3}$	=	30
×		÷		×
$\frac{1}{4}$	÷	$\frac{1}{9}$	=	$2\frac{1}{4}$
=		=		=
$4\frac{1}{2}$	×			

Final:

18	×	$1\frac{2}{3}$	=	30
×		÷		×
$\frac{1}{4}$	÷	$\frac{1}{9}$	=	$2\frac{1}{4}$
=		=		=
$4\frac{1}{2}$	×	15	=	$67\frac{1}{2}$

148. Step 1:

35	×	$\frac{3}{7}$	=	15
×		÷		×
$\frac{2}{5}$	÷	$\frac{1}{4}$	=	
=		=		=
14	×		=	

Step 2:

35	×	$\frac{3}{7}$	=	15
×		÷		×
$\frac{2}{5}$	÷	$\frac{1}{4}$	=	$1\frac{3}{5}$
=		=		=
14	×	$1\frac{5}{7}$	=	

Final:

35	×	$\frac{3}{7}$	=	15
×		÷		×
$\frac{2}{5}$	÷	$\frac{1}{4}$	=	$1\frac{3}{5}$
=		=		=
14	×	$1\frac{5}{7}$	=	24

149. Step 1:

$\frac{2}{3}$	×	5	=	
÷		÷		÷
$\frac{1}{36}$	×	4	=	$\frac{1}{9}$
=		=		=
24	×	$1\frac{1}{4}$	=	

Final:

$\frac{2}{3}$	×	5	=	$3\frac{1}{3}$
÷		÷		÷
$\frac{1}{36}$	×	4	=	$\frac{1}{9}$
=		=		=
24	×	$1\frac{1}{4}$	=	30

FRACTIONS

Fraction Fill 34-36

For each puzzle that follows, we are done when every square in the grid has been shaded or circled (to tell us it is unshaded). You may have used different steps or clues to help us arrive at the same final shading.

150. $1=\frac{4}{4}$, so the 1 in the bottom-right corner tells us that $\frac{4}{4}$ of its neighborhood cells are shaded. We shade all four cells.

$\frac{3}{4}$	$\frac{2}{3}$	$\frac{3}{4}$
$\frac{2}{3}$	$\frac{2}{3}$	$\frac{5}{6}$
$\frac{1}{2}$	$\frac{2}{3}$	1

$\frac{1}{2}$ of 4 is 2, so the $\frac{1}{2}$ in the bottom-left corner tells us that $\frac{2}{4}$ of its neighborhood cells are shaded.

Previously, we shaded two of the four cells. So, the two neighborhood cells that remain must be left unshaded.

$\frac{3}{4}$	$\frac{2}{3}$	$\frac{3}{4}$
$\left(\frac{2}{3}\right)$	$\frac{2}{3}$	$\frac{5}{6}$
$\left(\frac{1}{2}\right)$	$\frac{2}{3}$	1

Then, the $\frac{3}{4}$ in the top-left corner tells us that $\frac{3}{4}$ of its neighborhood cells are shaded. Previously, we found that 1 of these 4 cells cannot be shaded, so the three neighborhood cells that remain must all be shaded.

$\frac{3}{4}$	$\frac{2}{3}$	$\frac{3}{4}$
$\left(\frac{2}{3}\right)$	$\frac{2}{3}$	$\frac{5}{6}$
$\left(\frac{1}{2}\right)$	$\frac{2}{3}$	1

The $\frac{3}{4}$ in the top-right corner tells us that $\frac{3}{4}$ of its neighborhood cells are shaded. Previously, we shaded three of these cells. So, the neighborhood cell that remains must be left unshaded.

$\frac{3}{4}$	$\frac{2}{3}$	$\left(\frac{3}{4}\right)$
$\frac{2}{3}$	$\frac{2}{3}$	$\frac{5}{6}$
$\frac{1}{2}$	$\frac{2}{3}$	1

151. $1=\frac{4}{4}$, so the 1 in the top-right corner tells us that $\frac{4}{4}$ of its neighborhood cells are shaded. We shade all four cells.

$\frac{3}{4}$	$\frac{5}{6}$	1
$\frac{1}{2}$	$\frac{5}{9}$	$\frac{2}{3}$
$\frac{1}{2}$	$\frac{1}{2}$	$\frac{1}{2}$

$\frac{2}{3}$ of 6 is 4, so the $\frac{2}{3}$ indicated in the middle row tells us that $\frac{4}{6}$ of its neighborhood cells are shaded.

Since we have already shaded 4 of these 6 cells, the two cells that remain must be left unshaded.

$\frac{3}{4}$	$\frac{5}{6}$	1
$\frac{1}{2}$	$\frac{5}{9}$	$\frac{2}{3}$
$\frac{1}{2}$	$\left(\frac{1}{2}\right)$	$\left(\frac{1}{2}\right)$

The $\frac{5}{9}$ indicated in the middle row tells us that $\frac{5}{9}$ of its neighborhood cells are shaded.

There are only 9 cells in the entire grid, and we have already shaded 4. So, there is just one more shaded cell in the puzzle.

The only way we can satisfy all of the clues in the left column with just one more shaded cell is to shade the $\frac{1}{2}$ in the left column, middle row, as shown.

Then, the two remaining corner cells must be left unshaded.

152. $1 = \frac{4}{4}$, so the 1 in the bottom-right corner tells us that $\frac{4}{4}$ of its neighborhood cells are shaded. We shade all four cells.

$\frac{1}{2}$ of 4 is 2, so the $\frac{1}{2}$ in the top-right corner tells us that $\frac{2}{4}$ of its neighborhood cells are shaded. Since we have already shaded 2 of these 4 cells, the two neighborhood cells that remain must be left unshaded.

$\frac{2}{3}$ of 6 is 4, so the $\frac{2}{3}$ indicated in the top row tells us that $\frac{4}{6}$ of its neighborhood cells are shaded. Previously, we shaded 2 of these 6 cells. We also found that 2 of these 6 cells cannot be shaded. So, the two neighborhood cells that remain must both be shaded.

The $\frac{1}{4}$ in the bottom-left corner tells us that $\frac{1}{4}$ of its neighborhood cells are shaded. Since we have already shaded 1 of these 4 cells, the three neighborhood cells that remain must be left unshaded.

The $\frac{3}{4}$ in the top-left corner tells us that $\frac{3}{4}$ of its neighborhood cells are shaded. We found that one of these four cells cannot be shaded. So, the three other neighborhood cells are all shaded.

153.

Step 1:

$\frac{1}{2}$	$\frac{2}{3}$	$\frac{2}{3}$	$\frac{3}{4}$
$\frac{1}{3}$	$\frac{5}{9}$	$\frac{5}{9}$	$\frac{2}{3}$
0	$\frac{1}{3}$	$\frac{1}{2}$	$\frac{3}{4}$

Step 2:

$\frac{1}{2}$	$\frac{2}{3}$	$\frac{2}{3}$	$\frac{3}{4}$
$\frac{1}{3}$	$\frac{5}{9}$	$\frac{5}{9}$	$\frac{2}{3}$
0	$\frac{1}{3}$	$\frac{1}{2}$	$\frac{3}{4}$

Step 3:

$\frac{1}{2}$	$\frac{2}{3}$	$\frac{2}{3}$	$\frac{3}{4}$
$\frac{1}{3}$	$\frac{5}{9}$	$\frac{5}{9}$	$\frac{2}{3}$
0	$\frac{1}{3}$	$\frac{1}{2}$	$\frac{3}{4}$

Only 1 of these three

Step 4:

$\frac{1}{2}$	$\frac{2}{3}$	$\frac{2}{3}$	$\frac{3}{4}$
$\frac{1}{3}$	$\frac{5}{9}$	$\frac{5}{9}$	$\frac{2}{3}$
0	$\frac{1}{3}$	$\frac{1}{2}$	$\frac{3}{4}$

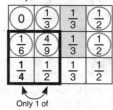

Only the middle square satisfies all of the clues.

Final:

$\frac{1}{2}$	$\frac{2}{3}$	$\frac{2}{3}$	$\frac{3}{4}$
$\frac{1}{3}$	$\frac{5}{9}$	$\frac{5}{9}$	$\frac{2}{3}$
0	$\frac{1}{3}$	$\frac{1}{2}$	$\frac{3}{4}$

154.

Step 1:

0	$\frac{1}{3}$	$\frac{1}{3}$	$\frac{1}{2}$
$\frac{1}{6}$	$\frac{4}{9}$	$\frac{1}{3}$	$\frac{1}{2}$
$\frac{1}{4}$	$\frac{1}{2}$	$\frac{1}{3}$	$\frac{1}{2}$

Step 2:

0	$\frac{1}{3}$	$\frac{1}{3}$	$\frac{1}{2}$
$\frac{1}{6}$	$\frac{4}{9}$	$\frac{1}{3}$	$\frac{1}{2}$
$\frac{1}{4}$	$\frac{1}{2}$	$\frac{1}{3}$	$\frac{1}{2}$

Step 3:

0	$\frac{1}{3}$	$\frac{1}{3}$	$\frac{1}{2}$
$\frac{1}{6}$	$\frac{4}{9}$	$\frac{1}{3}$	$\frac{1}{2}$
$\frac{1}{4}$	$\frac{1}{2}$	$\frac{1}{3}$	$\frac{1}{2}$

Step 4:

0	$\frac{1}{3}$	$\frac{1}{3}$	$\frac{1}{2}$
$\frac{1}{6}$	$\frac{4}{9}$	$\frac{1}{3}$	$\frac{1}{2}$
$\frac{1}{4}$	$\frac{1}{2}$	$\frac{1}{3}$	$\frac{1}{2}$

Only 1 of these two

Step 5:

0	$\frac{1}{3}$	$\frac{1}{3}$	$\frac{1}{2}$
$\frac{1}{6}$	$\frac{4}{9}$	$\frac{1}{3}$	$\frac{1}{2}$
$\frac{1}{4}$	$\frac{1}{2}$	$\frac{1}{3}$	$\frac{1}{2}$

Only 1 of these two

Step 6:

0	$\frac{1}{3}$	$\frac{1}{3}$	$\frac{1}{2}$
$\frac{1}{6}$	$\frac{4}{9}$	$\frac{1}{3}$	$\frac{1}{2}$
$\frac{1}{4}$	$\frac{1}{2}$	$\frac{1}{3}$	$\frac{1}{2}$

Final:

0	$\frac{1}{3}$	$\frac{1}{3}$	$\frac{1}{2}$
$\frac{1}{6}$	$\frac{4}{9}$	$\frac{1}{3}$	$\frac{1}{2}$
$\frac{1}{4}$	$\frac{1}{2}$	$\frac{1}{3}$	$\frac{1}{2}$

155.

Step 1:

3/4	2/3	2/3	3/4
5/6	2/3	2/3	2/3
1	2/3	2/3	1/2

Step 2:

3/4	2/3	2/3	3/4
5/6	2/3	(2/3)	2/3
1	**2/3**	(2/3)	1/2

Step 3:

3/4	2/3	2/3	3/4
5/6	2/3	(2/3)	2/3
1	2/3	**2/3**	1/2

Step 4:

3/4	2/3	2/3	**3/4**
5/6	2/3	(2/3)	2/3
1	2/3	(2/3)	1/2

Step 5:

3/4	(2/3)	**2/3**	3/4
5/6	2/3	(2/3)	2/3
1	2/3	(2/3)	1/2

Final:

3/4	(2/3)	2/3	3/4
5/6	2/3	(2/3)	2/3
1	2/3	(2/3)	1/2

156.

Step 1:

1/4	(1/6)	(0)	(1/3)	1/2
1/2	(4/9)	(1/3)	(4/9)	1/2
1/3	1/3	4/9	5/9	2/3
1/2	1/2	2/3	2/3	3/4

Step 2:

1/4	(1/6)	0	(1/3)	**1/2**
1/2	(4/9)	1/3	(4/9)	1/2
1/3	1/3	4/9	5/9	2/3
1/2	1/2	2/3	2/3	3/4

Step 3:

1/4	(1/6)	0	(1/3)	1/2
1/2	(4/9)	(**1/3**)	(4/9)	1/2
1/3	1/3	4/9	5/9	2/3
1/2	1/2	2/3	2/3	3/4

Step 4:

1/4	(1/6)	0	(1/3)	1/2
1/2	(4/9)	(1/3)	(4/9)	**1/2**
1/3	1/3	4/9	5/9	(2/3)
1/2	1/2	2/3	2/3	3/4

Step 5:

1/4	(1/6)	0	(1/3)	1/2
1/2	(4/9)	(1/3)	(4/9)	1/2
1/3	1/3	4/9	5/9	(2/3)
1/2	1/2	2/3	**2/3**	**3/4**

Step 6:

1/4	(1/6)	0	(1/3)	1/2
1/2	(4/9)	(1/3)	(4/9)	1/2
1/3	1/3	4/9	5/9	(2/3)
1/2	(1/2)	(**2/3**)	2/3	3/4

Step 7:

1/4	(1/6)	0	(1/3)	1/2
1/2	(4/9)	(1/3)	(4/9)	1/2
1/3	1/3	4/9	5/9	(2/3)
1/2	(1/2)	2/3	2/3	3/4

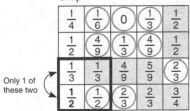

Only 1 of these two

Step 8:

1/4	(1/6)	0	(1/3)	1/2
(1/2)	(4/9)	(1/3)	(4/9)	1/2
1/3	(**1/3**)	4/9	5/9	(2/3)
1/2	(1/2)	(2/3)	2/3	3/4

Only 1 of these two

Step 9:

1/4	(1/6)	0	(1/3)	1/2
(1/2)	(4/9)	(1/3)	(4/9)	1/2
1/3	1/3	4/9	5/9	(2/3)
1/2	(1/2)	(2/3)	2/3	3/4

Step 10:

1/4	(1/6)	0	(1/3)	1/2
(**1/2**)	(4/9)	(1/3)	(4/9)	1/2
1/3	1/3	4/9	5/9	(2/3)
1/2	(1/2)	(2/3)	2/3	3/4

Final:

1/4	(1/6)	0	(1/3)	1/2
(1/2)	(4/9)	(1/3)	(4/9)	1/2
1/3	1/3	4/9	5/9	(2/3)
(**1/2**)	(1/2)	(2/3)	2/3	3/4

157.

Step 1:

1	**1**	1	2/3	1/2
2/3	7/9	8/9	2/3	1/2
1/2	5/9	7/9	2/3	2/3
1/4	1/3	2/3	2/3	3/4

Step 2:

1	1	1	**2/3**	1/2
2/3	7/9	8/9	2/3	1/2
1/2	5/9	7/9	2/3	2/3
1/4	1/3	2/3	2/3	3/4

Step 3:

1	1	1	2/3	(**1/2**)
2/3	7/9	8/9	2/3	(1/2)
1/2	5/9	7/9	2/3	2/3
1/4	1/3	2/3	2/3	3/4

Step 4:

1	1	1	2/3	(1/2)
2/3	7/9	8/9	2/3	(1/2)
(1/2)	(**5/9**)	7/9	2/3	2/3
1/4	1/3	2/3	2/3	3/4

Step 5:

1	1	1	2/3	(1/2)
2/3	7/9	**8/9**	2/3	(1/2)
1/2	(5/9)	7/9	2/3	2/3
1/4	1/3	2/3	2/3	3/4

Step 6:

1	1	1	2/3	(1/2)
2/3	7/9	8/9	2/3	(**1/2**)
(1/2)	(5/9)	7/9	2/3	(**2/3**)
1/4	1/3	2/3	2/3	3/4

Step 7:

1	1	1	2/3	(1/2)
2/3	7/9	8/9	2/3	(1/2)
(1/2)	(5/9)	7/9	2/3	(2/3)
1/4	1/3	2/3	**2/3**	**3/4**

Step 8:

1	1	1	2/3	(1/2)
2/3	7/9	8/9	2/3	(1/2)
(1/2)	(5/9)	7/9	2/3	(2/3)
1/4	1/3	(**2/3**)	**2/3**	3/4

Step 9:

1	1	1	2/3	(1/2)
2/3	7/9	8/9	2/3	(1/2)
(1/2)	(5/9)	7/9	2/3	(2/3)
1/4	**1/3**	(**2/3**)	2/3	3/4

Final:

1	1	1	2/3	(1/2)
2/3	7/9	8/9	2/3	(1/2)
(1/2)	(5/9)	7/9	2/3	(2/3)
(**1/4**)	1/3	(2/3)	2/3	3/4

158.

Step 1:

3/4	5/6	5/6	5/6	2/3	3/4
2/3	2/3	2/3	5/9	4/9	1/2
1/2	4/9	5/9	1/3	1/3	1/3
1/2	1/3	1/3	0	0	0

Step 2:

3/4	5/6	5/6	5/6	2/3	3/4
2/3	2/3	2/3	5/9	4/9	1/2
1/2	4/9	5/9	1/3	1/3	1/3
1/2	1/3	1/3	0	0	0

Step 3:

3/4	5/6	5/6	5/6	2/3	3/4
2/3	2/3	2/3	5/9	4/9	1/2
1/2	4/9	5/9	1/3	1/3	1/3
1/2	1/3	1/3	0	0	0

Step 4:

3/4	5/6	5/6	5/6	2/3	3/4
2/3	2/3	2/3	5/9	4/9	1/2
1/2	4/9	5/9	1/3	1/3	1/3
1/2	1/3	1/3	0	0	0

Step 5:

3/4	5/6	5/6	5/6	2/3	3/4
2/3	2/3	2/3	5/9	4/9	1/2
1/2	4/9	5/9	1/3	1/3	1/3
1/2	1/3	1/3	0	0	0

Step 6:

3/4	5/6	5/6	5/6	2/3	3/4
2/3	2/3	2/3	5/9	4/9	1/2
1/2	4/9	5/9	1/3	1/3	1/3
1/2	1/3	1/3	0	0	0

Step 7:

3/4	5/6	5/6	5/6	2/3	3/4
2/3	2/3	2/3	5/9	4/9	1/2
1/2	4/9	5/9	1/3	1/3	1/3
1/2	1/3	1/3	0	0	0

Step 8:

3/4	5/6	5/6	5/6	2/3	3/4
2/3	2/3	2/3	5/9	4/9	1/2
1/2	4/9	5/9	1/3	1/3	1/3
1/2	1/3	1/3	0	0	0

Step 9:

3/4	5/6	5/6	5/6	2/3	3/4
2/3	2/3	2/3	5/9	4/9	1/2
1/2	4/9	5/9	1/3	1/3	1/3
1/2	1/3	1/3	0	0	0

Final:

3/4	5/6	5/6	5/6	2/3	3/4
2/3	2/3	2/3	5/9	4/9	1/2
1/2	4/9	5/9	1/3	1/3	1/3
1/2	1/3	1/3	0	0	0

159.

Step 1:

1/2	1/3	0	1/3	1/2	3/4
1/2	4/9	1/3	5/9	2/3	5/6
2/3	2/3	5/9	5/9	2/3	5/6
3/4	5/6	5/6	2/3	2/3	3/4

Step 2:

1/2	1/3	0	1/3	1/2	3/4
1/2	4/9	1/3	5/9	2/3	5/6
2/3	2/3	5/9	5/9	2/3	5/6
3/4	5/6	5/6	2/3	2/3	3/4

Step 3:

1/2	1/3	0	1/3	1/2	3/4
1/2	4/9	1/3	5/9	2/3	5/6
2/3	2/3	5/9	5/9	2/3	5/6
3/4	5/6	5/6	2/3	2/3	3/4

Step 4:

1/2	1/3	0	1/3	1/2	3/4
1/2	4/9	1/3	5/9	2/3	5/6
2/3	2/3	5/9	5/9	2/3	5/6
3/4	5/6	5/6	2/3	2/3	3/4

160.

Step 1:

1/2	1/3	1/2	1/3	1/3	0
2/3	5/9	2/3	4/9	1/3	0
1	7/9	2/3	4/9	4/9	1/3
5/6	7/9	7/9	2/3	2/3	2/3
2/3	2/3	7/9	2/3	7/9	5/6
1/2	2/3	1	5/6	5/6	3/4

Step 2:

1/2	1/3	1/2	1/3	1/3	0
2/3	5/9	2/3	4/9	1/3	0
1	7/9	2/3	4/9	4/9	1/3
5/6	7/9	7/9	2/3	2/3	2/3
2/3	2/3	7/9	2/3	7/9	5/6
1/2	2/3	1	5/6	5/6	3/4

Step 3:

1/2	1/3	1/2	1/3	1/3	0
2/3	5/9	2/3	4/9	1/3	0
1	7/9	2/3	4/9	4/9	1/3
5/6	7/9	7/9	2/3	2/3	2/3
2/3	2/3	7/9	2/3	7/9	5/6
1/2	2/3	1	5/6	5/6	3/4

Step 4:

1/2	1/3	1/2	1/3	1/3	0
2/3	5/9	2/3	4/9	1/3	0
1	7/9	2/3	4/9	4/9	1/3
5/6	7/9	7/9	2/3	2/3	2/3
2/3	2/3	7/9	2/3	7/9	5/6
1/2	2/3	1	5/6	5/6	3/4

Step 5:

1/2	1/3	1/2	1/3	1/3	0
2/3	5/9	2/3	4/9	1/3	0
1	7/9	2/3	4/9	4/9	1/3
5/6	7/9	7/9	2/3	2/3	2/3
2/3	2/3	7/9	2/3	7/9	5/6
1/2	2/3	1	5/6	5/6	3/4

Step 6:

1/2	1/3	1/2	1/3	1/3	0
2/3	5/9	2/3	4/9	1/3	0
1	7/9	2/3	4/9	4/9	1/3
5/6	7/9	7/9	2/3	2/3	2/3
2/3	2/3	7/9	2/3	7/9	5/6
1/2	2/3	1	5/6	5/6	3/4

Step 7:

Step 8:

Step 9:

Step 10:

Step 11:

Final:

Step 5:

Step 6:

Step 7:

Step 8:

Step 9:

Step 10:

Step 11:

Step 12:

Step 13:

Step 14:

161. Step 1:

Step 2:

Step 3:

Step 4:

Step 15:

$\frac{3}{4}$	$\frac{2}{3}$	$\frac{1}{2}$	$\frac{1}{2}$	$\frac{1}{2}$	$\frac{1}{2}$
$\frac{2}{3}$	$\frac{5}{9}$	$\frac{4}{9}$	$\frac{1}{3}$	$\frac{4}{9}$	$\frac{1}{2}$
$\frac{5}{6}$	$\frac{2}{3}$	$\frac{5}{9}$	$\frac{1}{3}$	$\frac{4}{9}$	$\frac{1}{2}$
$\frac{2}{3}$	$\frac{4}{9}$	$\frac{2}{9}$	0	$\frac{1}{3}$	$\frac{1}{2}$
$\frac{5}{6}$	$\frac{2}{3}$	$\frac{4}{9}$	$\frac{2}{9}$	$\frac{1}{3}$	$\frac{1}{3}$
$\frac{3}{4}$	$\frac{2}{3}$	$\frac{1}{2}$	$\frac{1}{3}$	$\frac{1}{3}$	$\frac{1}{4}$

Final:

$\frac{3}{4}$	$\frac{2}{3}$	$\frac{1}{2}$	$\frac{1}{2}$	$\frac{1}{2}$	$\frac{1}{2}$
$\frac{2}{3}$	$\frac{5}{9}$	$\frac{4}{9}$	$\frac{1}{3}$	$\frac{4}{9}$	$\frac{1}{2}$
$\frac{5}{6}$	$\frac{2}{3}$	$\frac{5}{9}$	$\frac{1}{3}$	$\frac{4}{9}$	$\frac{1}{2}$
$\frac{2}{3}$	$\frac{4}{9}$	$\frac{2}{9}$	0	$\frac{1}{3}$	$\frac{1}{2}$
$\frac{5}{6}$	$\frac{2}{3}$	$\frac{4}{9}$	$\frac{2}{9}$	$\frac{1}{3}$	$\frac{1}{3}$
$\frac{3}{4}$	$\frac{2}{3}$	$\frac{1}{2}$	$\frac{1}{3}$	$\frac{1}{3}$	$\frac{1}{4}$

166. Dividing by a number is the same as multiplying by the reciprocal of that number. The reciprocal of any unit fraction $\frac{1}{n}$ is the whole number n.

So, **dividing a whole number by a unit fraction is the same as multiplying a whole number by another whole number. This will always give a whole number result.**

FRACTIONS

Challenge 37

162. We use k to represent Kyle's favorite number and write an equation: $k \times 35 = 50$. To find the value of k, we use the relationship between multiplication and division: if $k \times 35 = 50$, then $50 \div 35 = k$.

So, Kyle's favorite number is $50 \div 35 = \frac{50}{35} = \frac{10}{7} = 1\frac{3}{7}$.

163. The equation $\square \div \frac{1}{5} = 70$ means the same as $\square \times 5 = 70$.
If $\square \times 5 = 70$, then $70 \div 5 = \square$.

$70 \div 5 = 14$, so **14** is the number that gives quotient 70 when divided by $\frac{1}{5}$.
We check that $14 \div \frac{1}{5} = 14 \times 5 = 70$. ✓

164. We check the number of cakes we can make from each of the ingredients.

<u>Eggs:</u> Each cake requires 3 eggs, so 50 eggs allows us to make $50 \div 3 = \frac{50}{3} = 16\frac{2}{3}$ cakes.

<u>Cocoa:</u> Each cake requires $\frac{1}{3}$ cups of cocoa, so 6 cups of cocoa allows us to make $6 \div \frac{1}{3} = 6 \times 3 = 18$ cakes.

<u>Sugar:</u> Each cake requires $\frac{1}{2}$ cups of sugar, so 7 cups of sugar allows us to make $7 \div \frac{1}{2} = 7 \times 2 = 14$ cakes.

<u>Butter:</u> Each cake requires $\frac{1}{4}$ cups of butter, so 8 cups of butter allows us to make $8 \div \frac{1}{4} = 8 \times 4 = 32$ cakes.

The smallest number above is 14, which is the greatest number of cakes Laura can make with all of her sugar. If she tries to make more than 14 cakes, she won't have enough sugar. So, with only the ingredients listed, Laura can make at most **14** cakes.

165. We draw a diagram to represent Billy's favorite number. When Billy adds $2\frac{1}{2}$ to $\frac{4}{5}$ of his favorite number, he gets his favorite number.

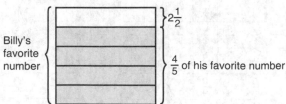

Billy's favorite number

$2\frac{1}{2}$

$\frac{4}{5}$ of his favorite number

Now, we can see that $2\frac{1}{2}$ is equal to $\frac{1}{5}$ of Billy's favorite number. So, Billy's favorite number is $2\frac{1}{2} \times 5 = \mathbf{12\frac{1}{2}}$.

Place Value
pages 39-41

1. 0.7 has 0 ones and 7 tenths. So, $0.7 = \frac{7}{10}$.

2. 0.3 has 0 ones and 3 tenths. So, $0.3 = \frac{3}{10}$.

3. 0.04 has 0 ones, 0 tenths, and 4 hundredths. So, $0.04 = \frac{4}{100}$.

4. 0.05 has 0 ones, 0 tenths, and 5 hundredths. So, $0.05 = \frac{5}{100}$.

5. 0.01 has 0 ones, 0 tenths, and 1 hundredth. So, $0.01 = \frac{1}{100}$.

6. 0.008 has 0 ones, 0 tenths, 0 hundredths, and 8 thousandths. So, $0.008 = \frac{8}{1,000}$.

7. 0.009 has 0 ones, 0 tenths, 0 hundredths, and 9 thousandths. So, $0.009 = \frac{9}{1,000}$.

8. 0.006 has 0 ones, 0 tenths, 0 hundredths, and 6 thousandths. So, $0.006 = \frac{6}{1,000}$.

9. 0.08 has 0 ones, 0 tenths, and 8 hundredths. So, $0.08 = \frac{8}{100}$.

10. 0.0003 has 0 ones, 0 tenths, 0 hundredths, 0 thousandths, and 3 ten-thousandths. So, $0.0003 = \frac{3}{10,000}$.

11. The hundredths digit is the second digit to the right of the decimal point. We circle this digit for each of the three decimal numbers.

945.8⑥   165.4⓪2   82.9④5

12. The hundredths digit is the second digit to the right of the decimal point. We insert decimal points as shown so that each number has a 5 two digits to the right of the decimal point.

30.4_5_   1.4_5_9   5.3_5_01

13. The tens digit is the second digit to the left of the decimal point. The tenths digit is the first digit to the right of the decimal point. We insert decimal points as shown so that each number has a 3 two digits to the left of the decimal point and a 4 one digit to the right of the decimal point.

3_2_.4_2_   3_4_.4_3_4   34_3_3._4_

14. The hundreds digit is the third digit to the left of the decimal point. The tenths digit is the first digit to the right of the decimal point. We insert decimal points as shown so that each number has the same hundreds and tenths digits.

7_8_76._8_   1_1_22._1_   50_5_4._0_4

15. The tenths digit is the first digit to the right of the decimal point. The thousandths digit is the third digit to the right of the decimal point. We insert decimal points as shown so that each number has a tenths digit that is less than its thousandths digit.

15._3_2_4_   8._6_87_6_   92._0_2_4_

16. The decimal 0.7 has a 7 in the tenths place. So, $0.7 = \frac{7}{10}$. We convert $\frac{7}{10}$ to an equivalent fraction with denominator 1,000 by multiplying the numerator and denominator by 100.

$$0.7 = \frac{7}{10} \xrightarrow{\times 100} \frac{700}{1,000}$$

So, $0.7 = \frac{7}{10} = \frac{700}{1,000}$.

17. $0.02 = \frac{2}{100} \xrightarrow{\times 10} \frac{20}{1,000}$.

18. $0.09 = \frac{9}{100} \xrightarrow{\times 10} \frac{90}{1,000}$.

19. 0.5 has a 5 in the tenths place. So, $0.5 = \frac{5}{10}$. We convert $\frac{5}{10}$ to an equivalent fraction with denominator 100 by multiplying the numerator and denominator by 10. We see that $\frac{5}{10} = \frac{50}{100}$. Then, we convert $\frac{50}{100}$ to an equivalent fraction with denominator 1,000 by multiplying the numerator and denominator by 10 again.

$$0.5 = \frac{50}{100} \xrightarrow{\times 10} \frac{500}{1,000}$$

20. $0.600 = \frac{6}{10} \xrightarrow{\times 100} \frac{600}{1,000}$.

21. $0.08 = \frac{8}{100} \xrightarrow{\times 10} \frac{80}{1,000}$.

22. $0.040 = \frac{4}{100} \xrightarrow{\times 10} \frac{40}{1,000}$.

23. $0.1 = \frac{1}{10} \xrightarrow{\times 10} \frac{10}{100}$.

24. We see that $\frac{30}{1,000} = \frac{3}{100}$. So, as a decimal $\frac{30}{1,000} = \mathbf{0.03}$. We can also write this as **0.030**.

Converting
42-44

25. $\frac{27}{100} = \frac{20}{100} + \frac{7}{100} = \frac{2}{10} + \frac{7}{100}$. So, to write $\frac{27}{100}$ as a decimal, we write 2 in the tenths place and 7 in the hundredths place: $\frac{27}{100} = \mathbf{0.27}$.

26. $\frac{291}{1,000} = \frac{200}{1,000} + \frac{90}{1,000} + \frac{1}{1,000} = \frac{2}{10} + \frac{9}{100} + \frac{1}{1,000} = \mathbf{0.291}$.

27. $7\frac{54}{100} = 7 + \frac{50}{100} + \frac{4}{100} = 7 + \frac{5}{10} + \frac{4}{100} = \mathbf{7.54}$.

28. $\frac{56}{1,000} = \frac{50}{1,000} + \frac{6}{1,000} = \frac{5}{100} + \frac{6}{1,000} = \textbf{0.056}$.

29. $0.21 = \frac{2}{10} + \frac{1}{100} = \frac{20}{100} + \frac{1}{100}$. So, to write 0.21 as a fraction, we add $\frac{20}{100} + \frac{1}{100} = \frac{21}{100}$.

30. $0.74 = \frac{7}{10} + \frac{4}{100} = \frac{70}{100} + \frac{4}{100} = \frac{74}{100}$.

31. $0.306 = \frac{3}{10} + \frac{0}{100} + \frac{6}{1,000} = \frac{300}{1,000} + \frac{0}{1,000} + \frac{6}{1,000} = \frac{306}{1,000}$.

32. $0.041 = \frac{4}{100} + \frac{1}{1,000} = \frac{40}{1,000} + \frac{1}{1,000} = \frac{41}{1,000}$.

33. In order to write 0.6050 as a fraction, we write
$0.6050 = \frac{6}{10} + \frac{5}{1,000} = \frac{600}{1,000} + \frac{5}{1,000} = \frac{605}{1,000}$.

34. 0.3 has one digit to the right of the decimal point, representing a number of tenths. So, $0.3 = \frac{3}{10}$.

35. 0.11 has two digits to the right of the decimal point, representing a number of hundredths. So, $0.11 = \frac{11}{100}$.

36. 0.029 has three digits to the right of the decimal point, representing a number of thousandths. So, $0.029 = \frac{29}{1,000}$.

37. 0.61 has two digits to the right of the decimal point, representing a number of hundredths. So, $0.61 = \frac{61}{100}$.

38. 0.444 has three digits to the right of the decimal point, representing a number of thousandths. So, $0.444 = \frac{444}{1,000}$.

39. 0.207 has three digits to the right of the decimal point, representing a number of thousandths. So, $0.207 = \frac{207}{1,000}$.

40. Since $\frac{89}{100}$ is a number of hundredths, the decimal equivalent has two digits after the decimal point. We write $\frac{89}{100} = \textbf{0.89}$.

41. Since $\frac{89}{1,000}$ is a number of thousandths, the decimal equivalent has three digits after the decimal point. We include a zero in the tenths place to write $\frac{89}{1,000} = \textbf{0.089}$.

42. Since $\frac{31}{100}$ is a number of hundredths, the decimal equivalent has two digits after the decimal point. We write $\frac{31}{100} = \textbf{0.31}$.

43. Since $\frac{31}{10,000}$ is a number of ten-thousandths, the decimal equivalent has four digits after the decimal point. We include zeros in the tenths and hundredths places to get $\frac{31}{10,000} = \textbf{0.0031}$.

44. $\frac{53}{100} = \textbf{0.53}$.

45. Since $\frac{9}{100} = 0.09$, we have $4\frac{9}{100} = \textbf{4.09}$.

46. We write $\frac{913}{100} = 9\frac{13}{100}$. Then, since $\frac{13}{100} = 0.13$, we have $\frac{913}{100} = 9\frac{13}{100} = \textbf{9.13}$.

47. We write $\frac{179}{100} = 1\frac{79}{100}$. Then, since $\frac{79}{100} = 0.79$, we have $\frac{179}{100} = 1\frac{79}{100} = \textbf{1.79}$.

48. $\frac{237}{1,000} = \textbf{0.237}$.

49. We write $\frac{1,013}{1,000} = 1\frac{13}{1,000}$. Then, since $\frac{13}{1,000} = 0.013$, we have $\frac{1,013}{1,000} = 1\frac{13}{1,000} = \textbf{1.013}$.

50. Since $0.1 = \frac{1}{10}$, we have $12.1 = 12 + \frac{1}{10} = \textbf{12}\frac{1}{10}$.

51. Since $0.01 = \frac{1}{100}$, we have $12.01 = 12 + \frac{1}{100} = \textbf{12}\frac{1}{100}$.

52. Since $0.121 = \frac{121}{1,000}$, we have $1.121 = 1 + \frac{121}{1,000} = \textbf{1}\frac{121}{1,000}$.

53. Since $0.21 = \frac{21}{100}$, we have $11.21 = 11 + \frac{21}{100} = \textbf{11}\frac{21}{100}$.

54. Since $0.043 = \frac{43}{1,000}$, we have $2.043 = 2 + \frac{43}{1,000} = \textbf{2}\frac{43}{1,000}$.

55. Since $0.3 = \frac{3}{10}$, we have $234.3 = 234 + \frac{3}{10} = \textbf{234}\frac{3}{10}$.

56. As a decimal, "ninety-six hundredths" is written:
$\frac{96}{100} = \textbf{0.96}$.

57. Since $0.45 = \frac{45}{100}$, we have $1.45 = 1\frac{45}{100}$. We simplify this to $1\frac{45}{100} = \textbf{1}\frac{9}{20}$.

58. Since $0.075 = \frac{75}{1,000}$, we have $9.075 = 9\frac{75}{1,000}$. We simplify this to $9\frac{75}{1,000} = \textbf{9}\frac{3}{40}$.

59. We write all four numbers as mixed numbers in simplest form.

$1.087 = 1\frac{87}{1,000}$.

$\frac{187}{100} = 1\frac{87}{100}$.

one and eighty-seven hundredths $= 1\frac{87}{100}$.

$\frac{1,870}{1,000} = \frac{187}{100} = 1\frac{87}{100}$.

So, **1.087** is not equal to the other three.

$\boxed{1.087}$ $\quad$ $\frac{187}{100}$ $\quad$ one and eighty-seven hundredths $\quad$ $\frac{1,870}{1,000}$

60. From left to right, we convert each of the decimals in the list to fractions in simplest form:

$0.38 = \frac{38}{100} = \frac{19}{50}$ $\qquad$ $0.35 = \frac{35}{100} = \frac{7}{20}$ $\qquad$ $0.375 = \frac{375}{1,000} = \frac{3}{8}$

$0.385 = \frac{385}{1,000} = \frac{77}{200}$ $\qquad$ $3.8 = 3\frac{8}{10} = 3\frac{4}{5}$

Since $0.375 = \frac{3}{8}$, we circle our answer:

$\quad$ 0.38 $\quad$ 0.35 $\quad$ $\boxed{0.375}$ $\quad$ 0.385 $\quad$ 3.8

— or —

$8 \times 125 = 1,000$, so

$$\frac{3}{8} \xrightarrow{\times 125} \frac{375}{1,000} = 0.375.$$

We circle our answer:

$\quad$ 0.38 $\quad$ 0.35 $\quad$ $\boxed{0.375}$ $\quad$ 0.385 $\quad$ 3.8

61. We look at several fractions with numerator 45,078 whose denominator is a power of 10:

$\frac{45,078}{10} = 4,507.8$, $\quad$ $\frac{45,078}{100} = 450.78$, $\quad$ $\frac{45,078}{1,000} = 45.078$.

Dividing 45,078 by any power of 10 changes the placement of the decimal point without changing the order of the digits.

The only answer choice whose digits are all in the same order as 45,078 is **4.5078**. We circle our answer:

405.78   45.780   0.4578   (4.5078)   4.5780

Decimal Numberlink   47-49

62.
$\frac{7}{10} = 0.7$

$\frac{7}{100} = 0.07$

$\frac{77}{100} = 0.77$

$7\frac{7}{100} = 7.07$

$7\frac{77}{100} = 7.77$

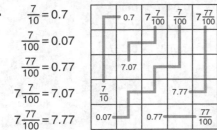

63.
$\frac{10}{100} = 0.1$

$\frac{11}{100} = 0.11$

$\frac{101}{100} = 1.01$

$\frac{1}{100} = 0.01$

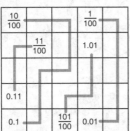

64.
$\frac{99}{100} = 0.99$

$\frac{999}{10} = 99.9$

$9\frac{99}{100} = 9.99$

$\frac{99}{10} = 9.9$

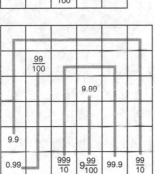

65.
$\frac{17}{10} = 1.7$

$\frac{107}{100} = 1.07$

$\frac{117}{10} = 11.7$

$\frac{17}{100} = 0.17$

$1\frac{17}{100} = 1.17$

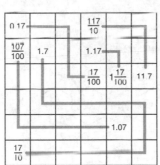

66.
$17\frac{76}{100} = 17.76$

$\frac{1,492}{1,000} = 1.492$

$\frac{1,776}{1,000} = 1.776$

$\frac{1,776}{10} = 177.6$

$\frac{1,492}{100} = 14.92$

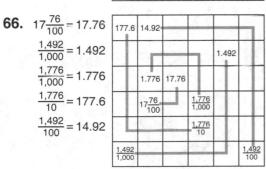

67.
$\frac{314}{100} = 3.14$

$\frac{314}{10} = 31.4$

$\frac{314}{1,000} = 0.314$

$\frac{341}{100} = 3.41$

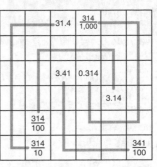

68.
$\frac{2}{10} = 0.2$

$\frac{22}{100} = 0.22$

$\frac{222}{100} = 2.22$

$\frac{2}{100} = 0.02$

$\frac{22}{10} = 2.2$

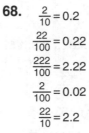

69.
$\frac{51}{10} = 5.1$

$\frac{15}{10} = 1.5$

$\frac{15}{100} = 0.15$

$\frac{115}{100} = 1.15$

$\frac{105}{100} = 1.05$

$\frac{501}{100} = 5.01$

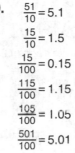

70.
$\frac{5}{10} = 0.50$

$\frac{55}{10} = 5.50$

$\frac{55}{100} = 0.55$

$\frac{5}{100} = 0.05$

$5\frac{5}{100} = 5.05$

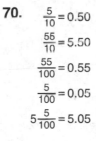

71.
$\frac{3,333}{100} = 33.33$

$\frac{333}{100} = 3.33$

$\frac{333}{10} = 33.3$

$\frac{33}{100} = 0.33$

$\frac{333}{1,000} = 0.333$

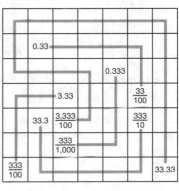

72. First, we write 0.4 = 0.400. Then, we write each of 0.400 and 0.789 as fractions:

$$0.400 = \frac{400}{1,000} \text{ and } 0.789 = \frac{789}{1,000}.$$

Since $\frac{400}{1,000} < \frac{789}{1,000}$, we have 0.4 $<$ 0.789.

73. First, we write 1.49 = 1.490. Then, we write each of 0.151 and 1.490 as fractions:

$$0.151 = \frac{151}{1,000} \text{ and } 1.490 = \frac{1,490}{1,000}.$$

Since $\frac{151}{1,000} < \frac{1,490}{1,000}$, we have 0.151 $<$ 1.49.

74. First, we write 1.53 = 1.530. Then, we write each of 1.524 and 1.530 as mixed numbers:

$$1.524 = 1\frac{524}{1,000} \text{ and } 1.530 = 1\frac{530}{1,000}.$$

Since $1\frac{524}{1,000} < 1\frac{530}{1,000}$, we have 1.524 $<$ 1.53.

75. First, we write 5.91 = 5.910. Then, we write each of 5.914 and 5.910 as mixed numbers:

$$5.914 = 5\frac{914}{1,000} \text{ and } 5.910 = 5\frac{910}{1,000}.$$

Since $5\frac{914}{1,000} > 5\frac{910}{1,000}$, we have 5.914 $>$ 5.91.

76. First, we write 0.64 = 0.640. Then, we write each of 0.064 and 0.640 as fractions:

$$0.064 = \frac{64}{1,000} \text{ and } 0.640 = \frac{640}{1,000}.$$

Since $\frac{64}{1,000} < \frac{640}{1,000}$, we have 0.064 $<$ 0.64.

77. We can write 7.003 = 7.0030. Then, we can write each of 7.0030 and 7.0003 as mixed numbers:

$$7.0030 = 7\frac{30}{10,000} \text{ and } 7.0003 = 7\frac{3}{10,000}.$$

Since $7\frac{30}{10,000} > 7\frac{3}{10,000}$, we have 7.003 $>$ 7.0003.

78. We write each decimal as a number of thousandths.

$$0.342 = \frac{342}{1,000} \qquad 0.243 = \frac{243}{1,000} \qquad 0.234 = \frac{234}{1,000}$$
$$0.432 = \frac{432}{1,000} \qquad 0.342 = \frac{342}{1,000} \qquad 0.423 = \frac{423}{1,000}$$

We see that $0.432 = \frac{432}{1,000}$ is the greatest number of thousandths, so **0.432** is the largest decimal.

79. We write each decimal as a number of thousandths:

$$0.02 = 0.020 = \frac{20}{1,000} \qquad 0.003 = \frac{3}{1,000}$$
$$0.3 = 0.300 = \frac{300}{1,000} \qquad 0.002 = \frac{2}{1,000}$$
$$0.03 = 0.030 = \frac{30}{1,000} \qquad 0.2 = 0.200 = \frac{200}{1,000}$$

We see that $0.002 = \frac{2}{1,000}$ is the smallest number of thousandths, so **0.002** is the smallest decimal.

80. Adding zeros to the end of a decimal number after the decimal point doesn't change its value. So, **0.78 = 0.780**.

(0.78) 0.078 0.708 0.807 (0.780) 0.87

81. Looking at the ones place, we see that 7.10 is the largest and 0.17 is the smallest. Each of 1.70 and 1.07 has a 1 in the ones place, so we look at the tenths place. Since 1.70 has more tenths than 1.07, we have 1.70>1.07. So, from greatest to least, we have:

$$7.10 > 1.70 > 1.07 > 0.17.$$

82. **0.49 > 0.409 > 0.0494 > 0.044.**

83. **1.21 > 1.12 > 1.102 > 1.021.**

84. **0.75 > 0.675 > 0.6705 > 0.657.**

85.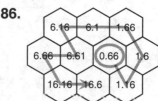

86.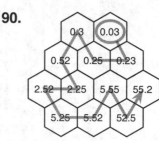

87.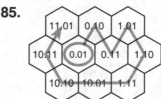

88.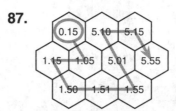

89.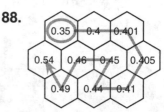

90.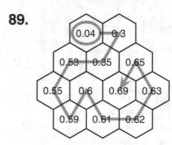

91.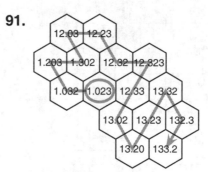

92.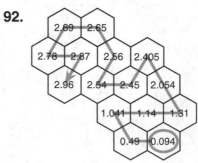

93.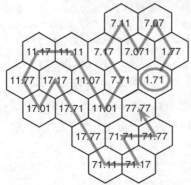

DECIMALS

Place Value & Ordering   54-55

94. Since all of the ones digits are blank and the greatest number cannot be more than 3.0, we use 1, 2, and 3 in the ones digits of the numbers from least to greatest. The remaining 3 fills the tenths place of the middle number.

$$1.4 < 2.3 < 3.0$$

95. The ones digit of the first two numbers must be 2 or less. So, the ones digits of the two smallest numbers cannot be 3 or 4. We fill these places with 1's:

$$1.3 < 1.\square < 2.1$$

Then, the tenths digit of the middle number must be greater than the 3 in the tenths digit of the smallest number. So, we place the 4 in the tenths place of the middle number. Then, we place the remaining 3 into the last empty box to create a true statement.

$$1.3 < 1.4 < 2.13$$

96. The ones digits of all three numbers are blank. We can fill them in with (3, 4, 4), (3, 4, 6), or (4, 4, 6).

If we place $3.5 < 4.\square < 4.2$, then we must place 6 in the remaining box. This makes an untrue statement.

If we place $4.5 < 4.\square < 6.2$, then we must place 3 in the remaining box. This makes an untrue statement.

So, we must fill in the ones digits as shown:

$$3.5 < 4.\square < 6.2$$

Then, we place the remaining 4 into the last empty box to create a true statement.

$$3.5 < 4.4 < 6.2$$

97. Since the ones digits of the first and last numbers are both 9, the ones digit of the middle number is also 9.

$$9.\square 8 < 9.7\square < 9.\square$$

Then, the number on the far right must have tenths digit larger than 7. The only remaining digit larger than 7 is 8.

$$9.\square 8 < 9.7\square < 9.8$$

Only placing the 5 and 7 into the remaining empty boxes as shown creates a true statement.

$$9.58 < 9.77 < 9.8$$

98. Since the ones digits of the third and fifth numbers are both 6, the ones digit of the fourth number must be a 6. Similarly, since the second number from the left has a 4 in the tenths place, the tenths digit of the leftmost number is at most 4. Since 4 is the smallest number we have available, we place the 4 in the tenths digit of the leftmost number:

$$5.45 < 5.4\square < 6 < 6.6 < 6.\square\square$$

Our remaining digits are 4, 5, 6, and 6. We see that the blank in the second number from the left must be greater than 5, so it must be 6. Similarly, the tenths digit of the rightmost number must be at least as large as the tenths digit of the second number from the right, so it must also be 6:

$$5.45 < 5.46 < 6 < 6.6\square < 6.6\square$$

Finally, we write the remaining 4 and 5 in increasing order as the hundredths digits of the last two numbers.

$$5.45 < 5.46 < 6 < 6.64 < 6.65$$

99. Since the rightmost number is less than 9 and the leftmost number is at least 6, the missing digits are all either 6's, 7's, or 8's. Of these, only **7** makes the statement true.

$$6.7 < 7.6 < 7.7 < 8.7$$

100. Since the rightmost number has ones digit 6, the units digit of the middle number cannot be more than 6. If we try 6, then we get 6.6<6.6, which is false. So, the digit must be less than 6.

Since the tenths digit of the smallest number is 4, the tenths digit of the middle number must be greater than 4.

So, only **5** makes the statement true.

$$5.45 < 5.5 < 6.5$$

101. Since the ones digits of the middle number is 5 and the hundredths digit of the middle number is less than 5, the missing digits must be less than 5. Then, since the hundredths digit of the middle number is 3, the missing digits must be greater than 3. Therefore, only **4** makes the statement true.

$$4.44 < 5.43 < 5.44$$

102. From the first statement, we know that the missing digits must be less than 4. Similarly, from the second statement, we know that the missing digit must be at least 3. Thus, the only possibility is **3**.

$$3.5 < 4.3 \ \& \ 3.3 < 3.4$$

103. Since $6.BC < 6.1C$, we know that B is less than 1. So, B is 0. Then, we have $5.08 < A.0C < 6.0C$. So, A must be at least 5, but less than 6. So, A is 5. Finally, we have $5.08 < 5.0C$. So, C must be greater than 8. This is only possible if C is 9. Therefore, $A = \mathbf{5}$, $B = \mathbf{0}$, and $C = \mathbf{9}$. Our final statement is

$$5.08 < 5.09 < 6.09 < 6.19$$

104. Since $3.JI < 3.KK$, we know that J is less than K.
Since $3.IJ < 3.JI$, we know that I is less than J.
Since $K.IJ < 3.IJ$, we know that K is less than 3.

So, we have $I<J<K<3$. Since I, J, and K are all different digits, this is only possible when $I=0$, $J=1$, and $K=2$. Our final statement is

$$2.01 < 3.01 < 3.10 < 3.22$$

105. Since $Z.Y6 < Z.YX$, we know that 6 is less than X.
Since $X.YZ < Z.Y6$, we know that X is less than Z.
Since $Z.YX < Y.ZX$, we know that Z is less than Y.

So, we have $6<X<Z<Y$. Since X, Y, and Z are all different digits, this is only possible when $X=7$, $Y=9$, and $Z=8$. Our final statement is

$$7.98 < 8.96 < 8.97 < 9.87$$

DECIMALS

Number Line 56-57

106. The medium tick marks split the number line between 7 and 8 into ten pieces. So, the medium tick marks represent tenths. The first blank is one tenth right of 7, so it is $7\frac{1}{10}$, or **7.1**. Similarly, the next blank is 6 tenths right of 7, at $7\frac{6}{10}$ or **7.6**.

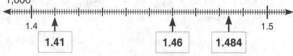

107. The medium tick marks split the number line between 1.4 and 1.5 into ten pieces. So, the medium tick marks represent hundredths. Similarly, the small tick marks divide each hundredth into thousandths.

The first blank is one hundredth right of 1.4. Since $1.4 = 1\frac{40}{100}$, the first blank is at $1\frac{41}{100}$, or **1.41**.

The next blank is 6 hundredths right of $1.4 = 1\frac{40}{100}$, at $1\frac{46}{100}$, or **1.46**.

The rightmost blank is 8 hundredths and 4 thousandths right of 1.4. Since $1.4 = 1\frac{400}{1,000}$, the rightmost blank is at $1\frac{484}{1,000}$, or **1.484**.

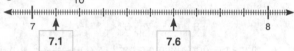

108. The medium tick marks split the number line between 2.52 and 2.56 into four hundredths. The small tick marks divide each hundredth into thousandths.

The first blank is two hundredths left of 2.52. Since $2.52 = 2\frac{52}{100}$, the first blank is at $2\frac{50}{100}$, or **2.5**.

The second blank is two hundredths and two thousandths right of 2.52. Since $2.52 = 2\frac{520}{1,000}$, the second blank is at $2\frac{542}{1,000}$, or **2.542**.

Finally, the rightmost blank is 4 hundredths right of $2.56 = 2\frac{56}{100}$, at $2\frac{60}{100}$, or **2.6**.

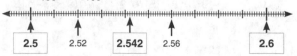

109. The medium tick marks split the number line between 7.05 and 7.06 into thousandths. The small tick marks split each thousandth into ten ten-thousandths.

The first blank is two thousandths right of 7.05. Since $7.05 = 7\frac{50}{1,000}$, the first blank is at $7\frac{52}{1,000}$, or **7.052**.

The second blank is two thousandths and nine ten-thousandths right of $7.052 = 7\frac{520}{10,000}$, at $7\frac{549}{10,000}$, or **7.0549**.

Finally, the rightmost blank is two thousandths left of $7.06 = 7\frac{60}{1,000}$, at $7\frac{58}{1,000}$, or **7.058**.

110. The medium tick marks split the number line between 9.9 and 10 into hundredths. The small tick marks split each hundredth into thousandths.

The first blank is three hundredths right of 9.9. Since $9.9 = 9\frac{90}{100}$, the first blank is at $9\frac{93}{100}$, or **9.93**.

The middle blank is two hundredths right of $9.93 = 9\frac{93}{100}$, at $9\frac{95}{100}$, or **9.95**.

The rightmost blank is 9 thousandths left of 10, at $9\frac{991}{1,000}$, or **9.991**.

111. The smallest tick marks split the number line between 5.022 and 5.079 into 57 thousandths. The second-smallest tick marks represent hundredths.

The leftmost blank is 2 hundredths and 2 thousandths left of 5.022. Since $5.022 = 5\frac{22}{1,000}$, the first blank is at **5**.
The middle blank is 20 thousandths right of $5.022 = 5\frac{22}{1,000}$, at $5\frac{42}{1,000}$, or **5.042**. Finally, the rightmost blank is 21 thousandths right of $5.079 = 5\frac{79}{1,000}$, at $5\frac{100}{1,000}$, or **5.1**.

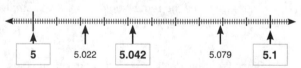

112. From 6 to 6.39 is 3 tenths and 9 hundredths. So, the medium tick marks represent tenths and the small tick marks represent hundredths. We label the numbers as shown below.

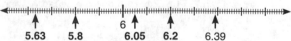

113. There are 30 hundredths between 6.3 and 6.6. So, the number exactly between them is $30\div2 = 15$ hundredths to the right of 6.3 and 15 hundredths to the left of 6.6, at **6.45**.

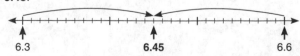

114. From least to greatest we have,

6.78 < 6.87 < 7.68 < 7.86 < 8.67 < 8.76.

115. Every number between 10 and 25 has two digits to the left of the decimal point. So, every number we can write between 10 and 25 using only the digits 1, 3, and 5 has a 1 in the tens place. This leaves us with 13.5 and 15.3. There are **2** different numbers.

116. Any number between 5.8 and 6.7 has one digit left of the decimal point, and that digit must be either a 5 or a 6. This gives us four possibilities: 5.67, 5.76, 6.57, and 6.75. Of these, only **6.57** is between 5.8 and 6.7.

117. The digits left of the decimal point must create a number greater than 4 and less than 25. So, each number we can write between 5 and 25 must be of the form 5.___, 23.__, or 24.__.

For 5.___, there are $3 \times 2 \times 1 = 6$ ways to arrange the remaining digits: 5.234, 5.243, 5.324, 5.342, 5.423, and 5.432.

For 23.__, there are $2 \times 1 = 2$ ways to arrange the remaining digits: 23.45, and 23.54.

For 24.__, there are $2 \times 1 = 2$ ways to arrange the remaining digits: 24.35, and 24.53.

So, in total we have $6 + 2 + 2 = $ **10** decimals.

118. Since $A > B$, we know $0.A$ has more tenths than $0.B$. So, we have $0.A \,\textcircled{>}\, 0.B$.

119. $0.0A$ and $0.B$ have the same number of ones, so we look at the tenths. Since $B > 0$, we know that $0.B$ has more tenths than $0.0A$. So, we have $0.0A \,\textcircled{<}\, 0.B$.

120. $B.AB$ and $B.B$ have the same number of ones, so we look at the tenths. Since $A > B$, we know $B.AB$ has more tenths than $B.B$. So, we have $B.B \,\textcircled{<}\, B.AB$.

121. $A.0A$ and $A.0B$ have the same number of ones and tenths, so we look at the hundredths. Since $A > B$, we know $A.0A$ has more hundredths than $A.0B$. So, we have $A.0A \,\textcircled{>}\, A.0B$.

122. Since $A > B$, we know $A.BB$ has more ones than $B.BA$. So, we have $A.BB \,\textcircled{>}\, B.BA$.

123. $0.00B$ and $0.0A$ have the same number of ones and tenths, so we look at the hundredths. Since $A > 0$, we know that $0.0A$ has more hundredths than $0.00B$. So, we have $0.00B \,\textcircled{<}\, 0.0A$.

124. $0.DE$ and $0.ED$ have the same number of ones, so we look at the tenths. Since $D > E$, we know $0.DE$ has more tenths than $0.ED$. So, we have $0.DE \,\textcircled{>}\, 0.ED$.

125. $0.E$ and $0.0D$ have the same number of ones, so we look at the tenths. Since $E > 0$, we know that $0.E$ has more tenths than $0.0D$. So, we have $0.E \,\textcircled{>}\, 0.0D$.

126. Since $D > 0$, we know that $DE.F$ has more tens than $D.EF$. So, we have $D.EF \,\textcircled{<}\, DE.F$.

127. Since adding a zero at the end of a decimal number doesn't change its value, we have $0.DF \,\textcircled{=}\, 0.DF0$.

128. $0.DF$ and $0.EE$ have the same number of ones, so we look at the tenths. Since $D > E$, we know $0.DF$ has more tenths than $0.EE$. So, we have $0.DF \,\textcircled{>}\, 0.EE$.

129. $0.0D$ and $0.0E$ have the same number of ones and tenths, so we look at the hundredths. Since $D > E$, we know $0.0D$ has more hundredths than $0.0E$, so $0.0D \,\textcircled{>}\, 0.0E$.

130. 1.92 is between 1.9 and 2.0. Since 1.92 is closer to 1.9, we round 1.92 down to **1.9**.

131. 4.86 is between 4.8 and 4.9. Since 4.86 is closer to 4.9, we round 4.86 up to **4.9**.

132. 7.01 is between 7.0 and 7.1. Since 7.01 is closer to 7.0, we round 7.01 down to **7.0**.

133. 13.956 is between 13.9 and 14.0. Since 13.956 is closer to 14.0, we round 13.956 up to **14.0**.

134. 9.873 is between 9.87 and 9.88. Since 9.873 is closer to 9.87, we round 9.873 down to **9.87**.

135. 5.075 is between 5.07 and 5.08. Since 5.075 is exactly halfway between 5.07 and 5.08, we round 5.075 up to **5.08**.

136. 44.444 is between 44.44 and 44.45. Since 44.444 is closer to 44.44, we round 44.444 down to **44.44**.

137. 49.9999 is between 49.99 and 50.00. Since 49.9999 is closer to 50.00, we round 49.9999 up to **50.00**.

138. 6.7892 is between 6.789 and 6.790. Since 6.7892 is closer to 6.789, we round 6.7892 down to **6.789**.

139. 1.8765 is between 1.876 and 1.877. Since 1.8765 is exactly halfway between 1.876 and 1.877, we round 1.8765 up to **1.877**.

140. 0.000451 is between 0.000 and 0.001. Since 0.000451 is closer to 0.000, we round 0.000451 down to **0.000**.

141. 0.0395 is between 0.039 and 0.040. Since 0.0395 is exactly halfway between 0.039 and 0.040, we round 0.0395 up to **0.040**.

142. For 0.4_5 to round to 0.45 when rounded to the nearest hundredth, the hundredths digit must be a four: 0.4<u>4</u>5. We check that this number satisfies the problem:

0.445 rounded to the nearest tenth is 0.4. ✓
0.445 rounded to the nearest hundredth is 0.45. ✓

143. The tenths digit is the first digit right of the decimal point. So, we can place the decimal point to the left of a 4 that will remain a 4 when the number is rounded, or to the left of a 3 that will become a 4 when the number is rounded.

Below, we show the only way to place each decimal point so that the tenths digit of each rounded number is 4.

3.4653   42.396   139.47.

We check our answers:
3.4653 rounds to 3.47. ✓
42.396 rounds to 42.40. ✓
139.47 rounds to 139.47. ✓

144. The smallest number that rounds to 5.00 must round up to 5.00. So, it must be between 4.99 and 5.00. When rounding to the nearest hundredth, in order to round up, the thousandths digit must be at least 5. So, the smallest number between 4.99 and 5.00 that rounds up to 5.00 is **4.995**.

145. Numbers that round to 5 when rounded to the nearest whole number are at least 4.5 but less than 5.5. We list the numbers that Alex writes in this range: 4.536, 4.563, 4.635, 4.653, 5.346, 5.364, 5.436, and 5.463. There are **8** numbers all together.

DECIMALS
Addition   62-63

146. We add the hundredths, then the tenths, then the ones. The decimal point goes between the ones and the tenths.

$$\begin{array}{r} 4.31 \\ + 5.28 \\ \hline 9 \end{array} \qquad \begin{array}{r} 4.31 \\ + 5.28 \\ \hline 59 \end{array} \qquad \begin{array}{r} 4.31 \\ + 5.28 \\ \hline \mathbf{9.59} \end{array}$$

147. First, we add the hundredths. 4+7 = 11 hundredths. We write a 1 in the hundredths place and regroup 10 hundredths to make 1 tenth, which we write above the 1 in the tenths place as shown.

$$\begin{array}{r} \scriptstyle 1 \\ 2.14 \\ + 7.37 \\ \hline 1 \end{array}$$

Then, we add the tenths and the ones.

$$\begin{array}{r} \scriptstyle 1 \\ 2.14 \\ + 7.37 \\ \hline 1 \end{array} \qquad \begin{array}{r} \scriptstyle 1 \\ 2.14 \\ + 7.37 \\ \hline 51 \end{array} \qquad \begin{array}{r} \scriptstyle 1 \\ 2.14 \\ + 7.37 \\ \hline \mathbf{9.51} \end{array}$$

148.
$$\begin{array}{r} \scriptstyle 1\ 1 \\ 9.57 \\ + 6.43 \\ \hline \mathbf{16.00} \end{array}$$

149. We write 0.69 as 0.690 so that both numbers have digits in the same place values. Then, we add.

$$\begin{array}{r} \scriptstyle 1\ 1 \\ 5.381 \\ + 0.690 \\ \hline \mathbf{6.071} \end{array}$$

150.
$$\begin{array}{r} \scriptstyle 1\ \ 1 \\ 9.909 \\ + 0.101 \\ \hline \mathbf{10.010} \end{array}$$

151.
$$\begin{array}{r} \scriptstyle 1\ 1\ 1 \\ 4.444 \\ + 0.888 \\ \hline \mathbf{5.332} \end{array}$$

152. We stack the decimals so that the sum is easier to compute.

$$\begin{array}{r} 9.21 \\ + 3.07 \\ \hline \mathbf{12.28} \end{array}$$

153.
$$\begin{array}{r} \scriptstyle 1\ 1 \\ 5.84 \\ + 3.19 \\ \hline \mathbf{9.03} \end{array}$$

154. We write 11.2 as 11.20 so that both numbers have digits in the same place values.

$$\begin{array}{r} 11.20 \\ + 0.43 \\ \hline \mathbf{11.63} \end{array}$$

155. We write 6.75 as 6.750 so that both numbers have digits in the same place values.

$$\begin{array}{r} \scriptstyle 1 \\ 6.750 \\ + 8.056 \\ \hline \mathbf{14.806} \end{array}$$

156. We write 0.3 as 0.300 and 0.55 as 0.550 so that all three numbers have digits in the same place values. Then, we stack all three numbers vertically and add them.

$$\begin{array}{r} \scriptstyle 1\ 1 \\ 0.300 \\ 0.550 \\ + 0.777 \\ \hline \mathbf{1.627} \end{array}$$

157. We write 6.6 as 6.600 and 0.77 as 0.770 so that all three numbers have digits in the same place values. Then, we stack all three numbers vertically and add them.

$$\begin{array}{r} \scriptstyle 1\ 1 \\ 6.600 \\ 0.770 \\ + 0.088 \\ \hline \mathbf{7.458} \end{array}$$

158. We write each fraction as a decimal, then add.
$\frac{1}{10} = 0.1$, $\frac{23}{100} = 0.23$, and $\frac{45}{1,000} = 0.045$.

We write zeros at the ends of 0.1 and 0.23 so that all three numbers have digits in the same place values. Then, we stack the decimals and add them as shown.

$$\begin{array}{r} 0.100 \\ 0.230 \\ + 0.045 \\ \hline \mathbf{0.375} \end{array}$$

159. We write zeros at the ends of 0.4 and 0.56 so that all three numbers have digits in the same place values. Then, we stack the decimals and add them as shown.

$$\begin{array}{r} \scriptstyle 1\ 1 \\ 0.400 \\ 0.560 \\ + 0.078 \\ \hline 1.038 \end{array}$$

Then, we write 1.038 as a mixed number and simplify:
$1.038 = 1\frac{38}{1,000} = 1\frac{19}{500}$.

160. The decimals Winnie writes that are less than 5.5 are 3.57, 3.75, and 5.37. We add them as shown below.

$$\begin{array}{r} \scriptstyle 1\ 1 \\ 3.57 \\ 3.75 \\ + 5.37 \\ \hline \mathbf{12.69} \end{array}$$

DECIMALS
Subtraction   64-65

161. We subtract the hundredths, then the tenths, then the ones as shown.

$$\begin{array}{r} 7.39 \\ - 3.26 \\ \hline 3 \end{array} \qquad \begin{array}{r} 7.39 \\ - 3.26 \\ \hline 13 \end{array} \qquad \begin{array}{r} 7.39 \\ - 3.26 \\ \hline \mathbf{4.13} \end{array}$$

162. We cannot take 7 hundredths from 4 hundredths, so we take 1 tenth from the tenths place of 8.34 and break it into 10 hundredths. This gives us 8 ones, 2 tenths, and 14 hundredths to subtract from.

$$\begin{array}{r} \overset{2\ 14}{8.\cancel{3}\cancel{4}} \\ -\ 2.17 \\ \hline \end{array}$$

Then, we subtract the hundredths, tenths, and ones.

$$\begin{array}{r} \overset{2\ 14}{8.\cancel{3}\cancel{4}} \\ -\ 2.17 \\ \hline 7 \end{array} \qquad \begin{array}{r} \overset{2\ 14}{8.\cancel{3}\cancel{4}} \\ -\ 2.17 \\ \hline 17 \end{array} \qquad \begin{array}{r} \overset{2\ 14}{8.\cancel{3}\cancel{4}} \\ -\ 2.17 \\ \hline \mathbf{6.17} \end{array}$$

163. We cannot take 8 hundredths from 7 hundredths. We don't have any tenths to take from the tenths place of 9.07. So, we take 1 one from the ones place of 9.07 and break it into 10 tenths.

$$\begin{array}{r} \overset{8\ 10}{\cancel{9}.\cancel{0}7} \\ -\ 4.18 \\ \hline \end{array}$$

Then, we take 1 of these tenths and break it into 10 hundredths. This gives us 8 ones, 9 tenths, and 17 hundredths to subtract from.

$$\begin{array}{r} \overset{9}{\overset{8\ 10\ 17}{\cancel{9}.\cancel{0}\cancel{7}}} \\ -\ 4.18 \\ \hline \end{array}$$

Then, we subtract the hundredths, tenths, and ones.

$$\begin{array}{r} \overset{9}{\overset{8\ 10\ 17}{\cancel{9}.\cancel{0}\cancel{7}}} \\ -\ 4.18 \\ \hline 9 \end{array} \qquad \begin{array}{r} \overset{9}{\overset{8\ 10\ 17}{\cancel{9}.\cancel{0}\cancel{7}}} \\ -\ 4.18 \\ \hline 89 \end{array} \qquad \begin{array}{r} \overset{9}{\overset{8\ 10\ 17}{\cancel{9}.\cancel{0}\cancel{7}}} \\ -\ 4.18 \\ \hline \mathbf{4.89} \end{array}$$

164. We write a zero at the end of 0.86 so that both numbers have digits in the same place values. We subtract the thousandths and hundredths.

$$\begin{array}{r} 5.481 \\ -\ 0.860 \\ \hline 21 \end{array}$$

We cannot take 8 tenths from 4 tenths. So, we take 1 one from the ones place of 5.481 and break it into 10 tenths. This leaves us with 4 ones and 14 tenths to subtract from.

$$\begin{array}{r} \overset{4\ 14}{\cancel{5}.\cancel{4}81} \\ -\ 0.860 \\ \hline 21 \end{array}$$

Then, we subtract the tenths and ones.

$$\begin{array}{r} \overset{4\ 14}{\cancel{5}.\cancel{4}81} \\ -\ 0.860 \\ \hline 621 \end{array} \qquad \begin{array}{r} \overset{4\ 14}{\cancel{5}.\cancel{4}81} \\ -\ 0.860 \\ \hline \mathbf{4.621} \end{array}$$

165. We can't take 9 thousandths away from 5 thousandths. We also don't have any hundredths or tenths to take from the hundredths or tenths places. So, we take 1 one from the ones place of 2.005 and break it into 10 tenths.

$$\begin{array}{r} \overset{1\ 10}{\cancel{2}.\cancel{0}05} \\ -\ 0.909 \\ \hline \end{array}$$

Then, we take 1 of those tenths and break it into 10 hundredths.

$$\begin{array}{r} \overset{9}{\overset{1\ 10\ 10}{\cancel{2}.\cancel{0}\cancel{0}5}} \\ -\ 0.909 \\ \hline \end{array}$$

Finally, we take 1 of those hundredths and break it into 10 thousandths. This gives us 1 one, 9 tenths, 9 hundredths, and 15 thousandths to subtract from.

$$\begin{array}{r} \overset{9\ 9}{\overset{1\ 10\ 10\ 15}{\cancel{2}.\cancel{0}\cancel{0}\cancel{5}}} \\ -\ 0.909 \\ \hline \end{array}$$

Then we subtract the thousandths, hundredths, tenths, and ones.

$$\begin{array}{r} \overset{9\ 9}{\overset{1\ 10\ 10\ 15}{\cancel{2}.\cancel{0}\cancel{0}\cancel{5}}} \\ -\ 0.909 \\ \hline 6 \end{array} \quad \begin{array}{r} \overset{9\ 9}{\overset{1\ 10\ 10\ 15}{\cancel{2}.\cancel{0}\cancel{0}\cancel{5}}} \\ -\ 0.909 \\ \hline 96 \end{array} \quad \begin{array}{r} \overset{9\ 9}{\overset{1\ 10\ 10\ 15}{\cancel{2}.\cancel{0}\cancel{0}\cancel{5}}} \\ -\ 0.909 \\ \hline 096 \end{array} \quad \begin{array}{r} \overset{9\ 9}{\overset{1\ 10\ 10\ 15}{\cancel{2}.\cancel{0}\cancel{0}\cancel{5}}} \\ -\ 0.909 \\ \hline \mathbf{1.096} \end{array}$$

166. We cannot take 7 thousandths from 5 thousands. So, we take 1 hundredth from the hundredths place of 5.555 and break it into 10 thousandths. This gives us 15 thousandths to subtract from.

$$\begin{array}{r} \overset{4\ 15}{5.5\cancel{5}\cancel{5}} \\ -\ 0.777 \\ \hline 8 \end{array}$$

We cannot take 7 hundredths from 4 hundredths. So, we take 1 tenth from the tenths place of 5.555 and break it into 10 hundredths. This gives us 14 hundredths to subtract from.

$$\begin{array}{r} \overset{4\ 14\ 15}{5.\cancel{5}\cancel{5}\cancel{5}} \\ -\ 0.777 \\ \hline 78 \end{array}$$

We cannot take 7 tenths from 4 tenths. So, we take 1 one from the ones place of 5.555 and break it into 10 tenths. This gives us 4 ones and 14 tenths to subtract from.

$$\begin{array}{r} \overset{4\ 14\ 14\ 15}{\cancel{5}.\cancel{5}\cancel{5}\cancel{5}} \\ -\ 0.777 \\ \hline .778 \end{array} \qquad \begin{array}{r} \overset{4\ 14\ 14\ 15}{\cancel{5}.\cancel{5}\cancel{5}\cancel{5}} \\ -\ 0.777 \\ \hline \mathbf{4.778} \end{array}$$

167. We write our problem vertically so the subtraction is easier to compute.

$$\begin{array}{r} 9.48 \\ -\ 3.07 \\ \hline \end{array}$$

Then, we subtract the hundredths, tenths, and ones.

$$\begin{array}{r} 9.48 \\ -\ 3.07 \\ \hline 1 \end{array} \qquad \begin{array}{r} 9.48 \\ -\ 3.07 \\ \hline .41 \end{array} \qquad \begin{array}{r} 9.48 \\ -\ 3.07 \\ \hline \mathbf{6.41} \end{array}$$

168. We cannot take 9 hundredths from 4 hundredths. So, we take 1 tenth from the tenths place of 9.84 and break it into 10 hundredths. This gives us 9 ones, 7 tenths, and 14 hundredths to subtract from.

$$\begin{array}{r} \overset{7\ 14}{9.\cancel{8}\cancel{4}} \\ -\ 3.29 \\ \hline \end{array}$$

Then, we subtract the hundredths, tenths, and ones.

$$\begin{array}{r} \overset{7\ 14}{9.\cancel{8}\cancel{4}} \\ -\ 3.29 \\ \hline 5 \end{array} \qquad \begin{array}{r} \overset{7\ 14}{9.\cancel{8}\cancel{4}} \\ -\ 3.29 \\ \hline 55 \end{array} \qquad \begin{array}{r} \overset{7\ 14}{9.\cancel{8}\cancel{4}} \\ -\ 3.29 \\ \hline \mathbf{6.55} \end{array}$$

169. We write 11.5 as 11.50 and 0.43 as 00.43 so that both numbers have digits in the same place values. We cannot take 3 hundredths from 0 hundredths. So, we take 1 tenth from the tenths place of 11.50 and break it into 10 hundredths. This leaves us with 1 ten, 1 one, 4 tenths, and 10 hundredths to subtract from.

$$\begin{array}{r} \overset{4\ 10}{11.\cancel{5}\cancel{0}} \\ -\ 00.43 \\ \hline \end{array}$$

Then, we subtract the hundredths, tenths, ones, and tens.

$$\begin{array}{r} \overset{4\ 10}{11.\cancel{5}\cancel{0}} \\ -\ 00.43 \\ \hline 7 \end{array} \quad \begin{array}{r} \overset{4\ 10}{11.\cancel{5}\cancel{0}} \\ -\ 00.43 \\ \hline 07 \end{array} \quad \begin{array}{r} \overset{4\ 10}{11.\cancel{5}\cancel{0}} \\ -\ 00.43 \\ \hline 1.07 \end{array} \quad \begin{array}{r} \overset{4\ 10}{11.\cancel{5}\cancel{0}} \\ -\ 00.43 \\ \hline \mathbf{11.07} \end{array}$$

170. We write 7.98 as 7.980 so that both numbers have digits in the same place values. We cannot take 3 thousandths from 0 thousandths. So, we take 1 hundredth from the hundredths place of 7.980 and break it into 10 thousandths. This leave us with with 7 ones, 9 tenths, 7 hundredths, and 10 thousandths to subtract from.

$$\begin{array}{r} \overset{7\ 10}{7.9\cancel{8}\cancel{0}} \\ -\ 4.023 \\ \hline \end{array}$$

Then, we subtract the thousandths, hundredths, tenths, and ones.

$$\begin{array}{r} \overset{7\ 10}{7.9\cancel{8}\cancel{0}} \\ -\ 4.023 \\ \hline 7 \end{array} \quad \begin{array}{r} \overset{7\ 10}{7.9\cancel{8}\cancel{0}} \\ -\ 4.023 \\ \hline 57 \end{array} \quad \begin{array}{r} \overset{7\ 10}{7.9\cancel{8}\cancel{0}} \\ -\ 4.023 \\ \hline 957 \end{array} \quad \begin{array}{r} \overset{7\ 10}{7.9\cancel{8}\cancel{0}} \\ -\ 4.023 \\ \hline \mathbf{3.957} \end{array}$$

171. We write 0.5 as 0.500 so that both numbers have digits in the same place values. We cannot take 9 thousandths from 0 thousandths. We don't have any hundredths to take from the hundredths place of 0.500. So, we take 1 tenth from the tenths place of 0.500 and break it into 10 hundredths.

$$\begin{array}{r} \overset{4\ 10}{0.\cancel{5}00} \\ -\ 0.009 \\ \hline \end{array}$$

Then, we take one of these hundredths and break it into 10 thousandths. This leaves us with 4 tenths, 9 hundredths, and 10 thousandths to subtract from.

$$\begin{array}{r} \overset{4\ \ 9}{0.\cancel{5}\overset{10}{\cancel{0}}0} \\ -\ 0.009 \\ \hline \end{array}$$

Then, we subtract the thousandths, hundredths, and

tenths.

$$\begin{array}{r} \overset{\ \ 9}{\overset{4\ 10}{0.\cancel{5}\cancel{0}0}} \\ -\ 0.009 \\ \hline 1 \end{array} \quad \begin{array}{r} \overset{\ \ 9}{\overset{4\ 10}{0.\cancel{5}\cancel{0}0}} \\ -\ 0.009 \\ \hline 91 \end{array} \quad \begin{array}{r} \overset{\ \ 9}{\overset{4\ 10}{0.\cancel{5}\cancel{0}0}} \\ -\ 0.009 \\ \hline \mathbf{0.491} \end{array}$$

172. We write 6.6 as 6.60 so that both numbers have digits in the same place values. We cannot take 7 hundredths from 0 hundredths so we take 1 tenth from the tenths place of 6.60 and break it into 10 hundredths. Then we subtract the hundredths.

$$\begin{array}{r} \overset{5\ 10}{6.\cancel{6}\cancel{0}} \\ -\ 0.77 \\ \hline 3 \end{array}$$

We cannot take 7 hundredths from 5 hundredths. So, we take 1 one from the ones place of 6.60 and break it into 10 tenths.

$$\begin{array}{r} \overset{\ \ 15}{\overset{5\ \cancel{6}\ 10}{\cancel{6}.\cancel{6}\cancel{0}}} \\ -\ 0.77 \\ \hline 3 \end{array}$$

Finally, we subtract the tenths and ones as shown.

$$\begin{array}{r} \overset{\ \ 15}{\overset{5\ \cancel{6}\ 10}{\cancel{6}.\cancel{6}\cancel{0}}} \\ -\ 0.77 \\ \hline \mathbf{5.83} \end{array}$$

173. We write each fraction as a decimal, then subtract. $\frac{12}{100} = 0.12$, and $\frac{56}{1,000} = 0.056$.

We write 0.12 as 0.120 so that both numbers have digits in the same place values. We stack the decimals and subtract them as shown:

$$\begin{array}{r} \overset{\ \ 11}{\overset{0\ 1\ 10}{0.\cancel{1}\cancel{2}0}} \\ -\ 0.056 \\ \hline \mathbf{0.064} \end{array}$$

174. First, we subtract as shown:

$$\begin{array}{r} \overset{\ \ 16}{\overset{5\ \cancel{6}\ 10}{0.\cancel{6}\cancel{7}0}} \\ -\ 0.089 \\ \hline 0.581 \end{array}$$

Then, we write 0.581 as a fraction: $\frac{581}{1,000}$. Since 581 and 1,000 do not have any factors in common, our answer is in simplest form.

175. We write each number as a decimal and subtract. "Forty-five and twenty-seven thousandths" is 45.027 and "thirty-one and nineteen hundredths" is 31.19. We align the numbers vertically and subtract as shown.

$$\begin{array}{r} \overset{\ \ \ 9}{\overset{4\ 10\ 12}{4\cancel{5}.\cancel{0}\cancel{2}7}} \\ -\ 31.190 \\ \hline \mathbf{13.837} \end{array}$$

DECIMALS
Arithmetic Strategies   66-67

176. Since $143 + 157 = 300$, we see that
$$0.143 + 0.157 = 0.300 = 0.3.$$
We reorder and regroup the terms and add:
$$(0.143 + 0.157) + 1.5 = 0.3 + 1.5 = \mathbf{1.8}.$$

177. We see that $0.22+0.78 = 1$ and $0.11+0.89 = 1$. So, we reorder and regroup the terms and add:
$$(0.89+0.11)+(0.78+0.22) = 1+1 = \textbf{2}.$$

178. Since $0.25+0.75 = 1$, we reorder and regroup the terms and add:
$$(0.25+0.75)+0.71 = 1+0.71 = \textbf{1.71}.$$

179. Since $1.09+0.41 = 1.5$, we rewrite our sum as
$$(1.09+0.41)+0.5 = 1.5+0.5 = \textbf{2.0 or 2}.$$

180. The perimeter of the triangle is the sum of the lengths of its sides. So, we compute the sum $0.053+0.04+0.027$. Since $0.053+0.027 = 0.08$, we write our sum as
$$(0.053+0.027)+0.04 = 0.08+0.04$$
$$= 0.12.$$

So, the triangle's perimeter is **0.12 cm**.

181. We count up. From 0.98 to 1 is 0.02. Then, from 1 to 6.52 is 5.52 more. So, the difference is $0.02+5.52 = \textbf{5.54}$.

– or –

Since 0.98 is close to 1, we add 0.02 to each of 6.52 and 0.98 to get $6.52-0.98 = 6.54-1 = \textbf{5.54}$.

182. We count up. From 2.88 to 3 is 0.12. Then, from 3 to 4.09 is 1.09 more. So, the difference is $0.12+1.09 = \textbf{1.21}$.

– or –

Since 2.88 is close to 3, we add 0.12 to each of 4.09 and 2.88 to get $4.09-2.88 = 4.21-3 = \textbf{1.21}$.

183. We count up. From 0.99 to 1 is 0.01. Then, from 1 to 1.97 is 0.97 more. So, the difference is $0.01+0.97 = \textbf{0.98}$.

– or –

Since 0.99 is close to 1, we add 0.01 to each of 1.97 and 0.99 to get $1.97-0.99 = 1.98-1 = \textbf{0.98}$.

184. We count up. From 1.84 to 2 is 0.16. Then, from 2 to 3.6 is 1.6 more. So, the difference is $0.16+1.6 = \textbf{1.76}$.

– or –

Since 1.84 is close to 2, we add 0.16 to each of 3.6 and 1.84 to get $3.6-1.84 = 3.76-2 = \textbf{1.76}$.

185. We first notice that $1.65+0.45 = 2.10$. So, we add $2.08+(1.65+0.45) = 2.08+2.10 = 4.18$.

Then, to subtract 0.93 from 4.18, we add 0.07 to each of 4.18 and 0.93 to make the subtraction easier:
$$4.18-0.93 = 4.25-1 = \textbf{3.25}.$$

– or –

To subtract 0.93 from 4.18, we count up. From 0.93 to 1 is 0.07. Then, from 1 to 4.18 is 3.18 more. So, the difference is $0.07+3.18 = \textbf{3.25}$.

186. Since $\frac{12}{100} = 0.12$ and $1.2 > 0.12$, we have $1.2 \; \boxed{>} \; \frac{12}{100}$.

187. Since $\frac{209}{1,000} = 0.209$ and $0.209 < 0.290$, we have $\frac{209}{1,000} \; \boxed{<} \; 0.290$.

188. Since $\frac{55}{100} = 0.55$ and $0.444 < 0.55$, we have $0.444 \; \boxed{<} \; \frac{55}{100}$.

189. We multiply the numerator and denominator of $\frac{23}{25}$ by 4 to see that this fraction is equivalent to $\frac{92}{100}$.

Since $\frac{92}{100} = 0.92$, we see that $0.92 \; \boxed{=} \; \frac{23}{25}$.

190. To find the perimeter, we add the lengths of the sides:
$$(4.07+1.93)+(4.07+1.93).$$

Since the sum in each pair of parentheses is 6, the perimeter is $6+6 = \textbf{12 cm}$.

191. To find the perimeter, we add the lengths of the sides:
$$(3.18+3.3)+(3.18+3.3).$$

Since the sum in each pair of parentheses is 6.48, the perimeter is $6.48+6.48 = \textbf{12.96 cm}$.

192. To find the perimeter, we add the lengths of the sides:
$$(2.42+0.878)+(2.42+0.878).$$

Since the sum in each pair of parentheses is 3.298, the perimeter is $3.298+3.298 = \textbf{6.596 in}$.

193. To find the perimeter, we add the lengths of the sides:
$$(1.2+1.573)+(1.2+1.573).$$

Since the sum in each pair of parentheses is 2.773, the perimeter is $2.773+2.773 = \textbf{5.546 in}$.

194. The distance Ray must run to finish the marathon is $26.2-17.25 = \textbf{8.95 miles}$.

195. Tok must lose $11.42-10.5 = \textbf{0.92 pounds}$.

196. The combined length of the first three pieces is
$$11.59+4.43+4.38 = 20.4 \text{ cm}.$$

Since the total length of the string is 25 cm, the fourth piece has length $25-20.4 = \textbf{4.6 cm}$.

197. We have $0.1 < 0.13 < 0.17 < 0.2 < 1.7 < 2.1$. Since 0.17 is between 0.13 and 0.2, these are the only two possibilities.

The distance from 0.13 to 0.17 is $0.17-0.13 = 0.04$. The distance from 0.2 to 0.17 is $0.2-0.17 = 0.03$.

Therefore, **0.2** is closest to 0.17.

0.1   0.13   (0.2)   1.7   2.1

198. We want to know what fraction 0.75 pounds is of 2.2 pounds. Since $0.75+0.75+0.75 = 2.25$, we see that 0.75 is about one-third of 2.2. We look at the choices to find a decimal that is close to one-third.

$$0.34 = \frac{34}{100}, \text{ which is very close to } \frac{33}{99} = \frac{1}{3}.$$

So, **0.34** is the best estimate.

199. Every number Shannon writes between 0 and 10 has one digit left of the decimal point. The largest number we can make is 6.54 and the smallest number is 4.56.

a. The sum of the smallest and largest numbers is $6.54 + 4.56 = \textbf{11.1}$.

b. Their difference is $6.54 - 4.56 = \textbf{1.98}$.

200. We start by stacking our addition problem as shown:

$$\begin{array}{r} A.BC \\ + \ C.BA \\ \hline 5.45 \end{array}$$

Looking at the hundredths digits above, we see that $C+A$ is either 5 or 15. But, looking at the ones digits, since $A.BC + C.BA$ is 5.45, neither A nor C can be greater than 5. So, $A+C = 5$.

Looking at the hundredths digits above, since $A+C = 5$ (not 15), there are no additional tenths to add in the tenths column. So, we have $B+B = 4$, which means $B = \textbf{2}$.

The ways A and C can sum to 5 are $5+0 = 5$, or $4+1 = 5$, or $3+2 = 5$. We are given that $A.BC - C.BA = 2.97$. This only works when $A = \textbf{4}$ and $C = \textbf{1}$.

We check our work.

$$\begin{array}{r} 4.21 \\ + \ 1.24 \\ \hline 5.45 \end{array} \qquad \begin{array}{r} 4.21 \\ - \ 1.24 \\ \hline 2.97 \end{array} \checkmark$$

201. The sum of all three rows is 18.9. Since $6.3 + 6.3 + 6.3 = 18.9$, the sum of the entries in each row is 6.3. So, the sum of the entries in each row, column, and diagonal is 6.3.

In the middle column, since $2.1 + 1.94 = 4.04$, the top cell in the middle column is $6.3 - 4.04 = \textbf{2.26}$.

Similarly, in the diagonal from bottom-left to top-right, since $2.1 + 2.23 = 4.33$, the bottom-left cell is $6.3 - 4.33 = \textbf{1.97}$.

	2.26	2.23
	2.1	
1.97	1.94	

We use the same method to fill the remaining cells as shown in the steps below:

1.81	2.26	2.23
	2.1	
1.97	1.94	2.39

1.81	2.26	2.23
2.52	2.1	1.68
1.97	1.94	2.39

202. $0.65 = \frac{65}{100}$. Dividing both the numerator and denominator by 5, we have $\frac{65}{100} = \frac{13}{20}$. So, $0.65 = \frac{\textbf{13}}{\textbf{20}}$.

203. $0.675 = \frac{675}{1,000}$. Dividing both the numerator and denominator by 25, we have $\frac{675}{1,000} = \frac{27}{40}$. So, $0.675 = \frac{\textbf{27}}{\textbf{40}}$.

204. $0.005 = \frac{5}{1,000}$. Dividing the numerator and denominator by 5, we have $\frac{5}{1,000} = \frac{1}{200}$. So, $0.005 = \frac{\textbf{1}}{\textbf{200}}$.

205. $0.936 = \frac{936}{1,000}$. Diving the numerator and denominator by 2, we have $\frac{936}{1,000} = \frac{468}{500}$. So, $0.936 = \frac{468}{500}$. Since the numerator and denominator are both still even, we divide by 2 again to see that $\frac{468}{500} = \frac{234}{250}$. So, $0.936 = \frac{234}{250}$. Again, the numerator and denominator are still even, so dividing by 2 again we see that $\frac{234}{250} = \frac{117}{125}$. Since 117 and 125 do not share any common factors besides 1, we know $\frac{117}{125}$ is in simplest form. So, $0.936 = \frac{\textbf{117}}{\textbf{125}}$.

206. Alex can write each of Grogg's decimals as a fraction with denominator 100 before simplifying. For example, $0.01 = \frac{1}{100}$, $0.28 = \frac{28}{100}$, and $0.99 = \frac{99}{100}$.

Simplifying each of these fractions gives a fraction whose denominator is a factor of 100.

Factors of 100 are 1, 2, 4, 5, 10, 20, 25, 50, and 100. We look for a fraction that gives each of these denominators.

<u>1</u>: Any fraction that simplifies to $\frac{\square}{1}$ is a whole number. Since Grogg's decimals are all between 0 and 1, none are whole numbers. So, 1 cannot be a denominator.

<u>2</u>: The only number less than 1 that simplifies to $\frac{\square}{2}$ is $\frac{1}{2} = \frac{50}{100} = 0.50$. But, Grogg does not write decimals with a 0 in the hundredths digit, so he does not write 0.50.

<u>4</u>: We have $\frac{1}{4} = \frac{25}{100} = 0.25$. So, 4 works. ✓

<u>5</u>: Numbers less than 1 that simplify to $\frac{\square}{5}$ are
$\frac{1}{5} = \frac{20}{100} = 0.20$, $\qquad \frac{2}{5} = \frac{40}{100} = 0.40$,
$\frac{3}{5} = \frac{60}{100} = 0.60$, $\quad$ and $\quad \frac{4}{5} = \frac{80}{100} = 0.80$.
Grogg does not write decimals with a 0 in the hundredths digit, so he does not write any of these.

<u>10</u>: Numbers less than 1 that simplify to $\frac{\square}{10}$ are
$\frac{1}{10} = 0.10$, $\frac{3}{10} = 0.30$, $\frac{7}{10} = 0.70$, and $\frac{9}{10} = 0.90$.
Grogg does not write decimals with a 0 in the hundredths digit, so he does not write any of these.

<u>20</u>: We have $\frac{1}{20} = \frac{5}{100} = 0.05$. So, 20 works. ✓

<u>25</u>: We have $\frac{1}{25} = \frac{4}{100} = 0.04$. So, 25 works. ✓

<u>50</u>: We have $\frac{1}{50} = \frac{2}{100} = 0.02$. So, 50 works. ✓

<u>100</u>: We have $\frac{1}{100} = 0.01$. So, 100 works. ✓

We found that 4, 20, 25, 50, and 100 are the possible denominators, for a total of **5** different denominators.

207. $0.ABC = \frac{ABC}{1,000}$. This fraction can only be simplified by canceling common factors in the numerator and denominator. So, when Grogg simplifies $\frac{ABC}{1,000}$, the resulting denominator must be a factor of 1,000. Since 7 is not a factor of 1,000, Grogg must have made a mistake.

PROBABILITY
Likelihood   Page 73

1. A person could be born in any one of the twelve months. So, it is **unlikely**, even **very unlikely** but not impossible, that the next stranger you will see has a birthday in April.

2. The next baby born in your town will be either a boy or a girl. So, it is almost **equally likely and unlikely** that the next baby born in your town will be a girl.
   In fact, there are slightly more boys born than girls. For every 100 girls born in the world, about 107 boys are born.

3. Cereal is a very common breakfast food. So, it is **very likely** that someone you know will have cereal for breakfast tomorrow. If you, or someone you know, eats cereal every day for breakfast, you might even say that this is **certain to happen**!

4. Walruses do not often deliver flowers to people. This silly scenario is **very unlikely** to happen. If you do not have a favorite uncle, you might even say this is **impossible**!

5. Roughly $\frac{3}{4}$ of the Earth's surface is covered in water, while the rest is land. So, it is **somewhat likely to very likely** that your finger is placed on a water region of a globe.

6. You've already thought a little about math today by working on this Beast Academy page! So, we can say it is **certain to happen** that you will think about math at least once today!

So, after connecting each event to its likelihood, your arrows should look similar to those shown below.

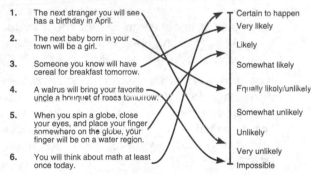

PROBABILITY
Counting Review   74-77

7. If we subtract 12 from each number in the list, we get a list of whole numbers from 1 to 224.

Now, the numbers in the list are counted for us! There are 224 numbers from 1 to 224.

Since this list is the same size as our original list, we know there are **224** numbers in the original list.

8. The numbers in this list count by threes, but they are not multiples of 3. To make the list easier to count, we first subtract 13 from each number in the list to make every number a multiple of 3:

$$3, 6, 9, \ldots, 48, 51, 54.$$

Then, we divide every number in this list by 3 to get

$$1, 2, 3, \ldots, 16, 17, 18.$$

So, there are **18** numbers in the original list.

9. We wish to count the even numbers between 15 and 65:

$$16, 18, 20, \ldots, 60, 62, 64.$$

We first divide every number in this list by 2 to get

$$8, 9, 10, \ldots, 30, 31, 32.$$

Then, we subtract 7 from every number in this list to get

$$1, 2, 3, \ldots, 23, 24, 25.$$

So, there are **25** even numbers between 15 and 65.

10. We wish to count the odd numbers between 10 and 200:

$$11, 13, 15, \ldots, 195, 197, 199.$$

We first subtract 9 from every number in the list to get

$$2, 4, 6, \ldots, 186, 188, 190.$$

Then, we divide every number in this list by 2 to get

$$1, 2, 3, \ldots, 93, 94, 95.$$

So, there are **95** odd numbers between 10 and 200.

11. We wish to count the multiples of 5 between 101 and 1,001:

$$105, 110, 115, \ldots, 990, 995, 1000.$$

We first subtract 100 from every number in the list to get

$$5, 10, 15, \ldots, 890, 895, 900.$$

Then, we divide every number in this list by 5 to get

$$1, 2, 3, \ldots, 178, 179, 180.$$

So, there are **180** multiples of 5 between 101 and 1,001.

12. There are three letters we could choose for the first spot in the arrangement. No matter which letter is chosen first, we have two remaining letters we could choose for the second spot. This leaves us with one remaining letter for the third spot.

All together, there are $3\times2\times1=$ **6** ways to arrange these three letters: ABC, ACB, BAC, BCA, CAB, and CBA.

— *or* —

There are three spots (first, second, or third) that we could place the A in. After placing the A, there are two remaining spots that we could choose for B, leaving one spot for C.

All together, there are $3\times2\times1=$ **6** ways to arrange these three letters: ABC, ACB, BAC, BCA, CAB, and CBA.

13. Every number that is divisible by 5 has units digit 0 or 5. So, if we arrange these digits into a five-digit multiple of 5, the units digit must be five: _ _,_ _ 5.

Then, the number is divisible by 5 for any arrangement of the other four digits. There are $4\times3\times2\times1=24$ ways to arrange the remaining four digits as the ten-thousands, thousands, hundreds, and tens digit of the number.

So, **24** arrangements of these digits create a 5-digit number that is divisible by 5.

14. Since Fred insists on sitting in the middle, there are 4 other chairs in which to seat Donna. Once Donna is seated, there are 3 possible chairs for Ed. Then, Grace will sit in one of the 2 remaining chairs. The 1 remaining chair must be filled by Hannah.

So, there are $4\times3\times2\times1=$ **24** ways to seat these five people if Fred sits in the middle.

— *or* —

Since Fred insists on sitting in the middle, there are 4 other people who could be seated in the first chair. Once that person is seated, there are 3 people that could be seated in the second chair. Then, one of the 2 remaining people will sit in the fourth chair. Finally, there is 1 remaining person who will fill the fifth chair.

So, there are $4\times3\times2\times1=$ **24** ways to seat these five people if Fred sits in the middle.

15. Even numbers are those with units digits 0, 2, 4, 6, or 8. So, if we arrange the given digits into a four-digit even number, we have two choices for a units digit:

$$\_ \, , \_ \_ \, 6 \quad or \quad \_ \, , \_ \_ \, 8.$$

Then, the number is divisible by 2 for any arrangement of the three other digits.

So, if the units digit is 6, then there are $3\times2\times1=6$ ways to arrange the remaining three digits as the thousands, hundreds, and tens digit of the number.

Similarly, if the units digit is 8, then there are $3\times2\times1=6$ ways to arrange the remaining three digits as the thousands, hundreds, and tens digit of the number.

So, $6+6=$ **12** arrangements of these digits create a four-digit number that is even.

16. We draw a tree diagram.

Die Roll	Coin Flip	
1	heads	**1.** 1H
	tails	**2.** 1T
2	heads	**3.** 2H
	tails	**4.** 2T
3	heads	**5.** 3H
	tails	**6.** 3T
4	heads	**7.** 4H
	tails	**8.** 4T
5	heads	**9.** 5H
	tails	**10.** 5T
6	heads	**11.** 6H
	tails	**12.** 6T

There are **12** different possible outcomes.

— *or* —

Winnie could roll 6 possible numbers on the die: 1, 2, 3, 4, 5, or 6.

For each possible die roll, there are 2 possible outcomes for the flip of a coin: heads or tails.

All together, there are $6\times2=$ **12** possible outcomes of rolling a die and flipping a coin.

17. Billy can choose any one of the 7 styles in any one of 3 colors. This gives him a total of $7\times3=$ **21** different choices for new frames.

If you drew a tree diagram, it might look like this:

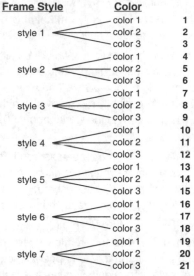

18. With 9 flavors and 13 topping choices, creating a tree diagram is not practical.

We can choose any one of the 13 toppings to go with any one of the 9 flavors of ice cream. This gives a total of $13\times9=$ **117** ways to make a cone with one flavor and one topping.

19. The odd digits are 1, 3, 5, 7, and 9. So, to make a four-digit number with only odd digits, we have 5 choices for each of the thousands, hundreds, tens and units digits. All together, this is $5\times5\times5\times5=$ **625** possible numbers.

20. To order a one-topping pizza, a customer must choose one of 3 sizes, one of 9 toppings, and one of 2 crust types. All together, this gives $3\times9\times2=$ **54** possible one-topping pizzas.

21. If we choose the fruit ingredients one at a time, there are 7 choices for the first fruit and 6 choices for the second fruit. However, the order in which we select the fruits doesn't matter! Since $7\times6=42$ counts each possible pair of fruits twice, there are $42\div2=$ **21** possible two-fruit smoothies.

22. If we choose the co-captains one at a time, there are 8 choices for the first co-captain and 7 choices for the second co-captain. However, the order in which we select the co-captains doesn't matter! Since $8\times7=56$ counts each possible pair of co-captains twice, there are $56\div2=$ **28** possible pairs of co-captains.

23. If we choose two girls one at a time, there are 5 choices for the first girl and 4 choices for the second girl. However, the order in which we select the girls doesn't matter.

Since $5 \times 4 = 20$ counts every pair of girls twice, there are $20 \div 2 = 10$ ways to pick a pair of girls from the five.

Similarly, there are 10 ways to pick a pair of boys from the five.

To form a group, we choose one of the 10 pairs of girls and one of the 10 pairs of boys. All together, we have $10 \times 10 = \textbf{100}$ ways to choose a group of two boys and two girls.

24. Twenty of the 150 jellybeans are black, and Katie is equally likely to select any of the jellybeans. So, the probability she chose a black jellybean is

$$\frac{\text{Ways to choose a black jellybean}}{\text{Total possible ways to choose a jellybean}} = \frac{20}{150} = \frac{\textbf{2}}{\textbf{15}}.$$

25. We first count the odd numbers from 1 through 75:

$$1, 3, 5, \ldots, 71, 73, 75.$$

Adding 1 to each number in this list, we get

$$2, 4, 6, \ldots, 72, 74, 76.$$

Then, dividing each number in the list by 2, we have

$$1, 2, 3, \ldots, 36, 37, 38.$$

So, there are 38 balls with odd numbers, and there are 75 possible balls we could draw. We are equally likely to draw any of these balls, so the probability that the number on the ball is odd is

$$\frac{\text{Ways to choose a ball with odd \#}}{\text{Total possible ways to choose a ball}} = \frac{\textbf{38}}{\textbf{75}}.$$

We notice this is a little more than $\frac{1}{2}$, which is what we expect because there are *almost* an equal number of balls with odd and even numbers.

26. Probability is greatest for an event that is certain to happen. For example, we are certain to roll a number from 1 to 6 on a standard six-sided die. When an event is certain to happen, the number of desired outcomes equals the number of possible outcomes.

When the number of desired outcomes is equal to the number of possible outcomes, the probability of that event is

$$\frac{\text{Number of desired outcomes}}{\text{Number of possible outcomes}} = \frac{\text{Number of possible outcomes}}{\text{Number of possible outcomes}} = 1.$$

Probability is smallest for an event that is impossible. For example, it is impossible to roll an 8 on a standard six-sided die. An event is impossible when *no* desired outcomes are possible outcomes.

When no desired outcomes are possible, the probability of that event is

$$\frac{\text{Number of desired outcomes}}{\text{Number of possible outcomes}} = \frac{0}{\text{Number of possible outcomes}} = 0.$$

So, the largest a probability could be is 1. The smallest a probability could be is 0.

Every probability is 0, 1, or between 0 and 1. We can use this fact to check that our answer to a probability problem is reasonable.

27. Thirteen of the cards are spades, and there are 52 possible cards we could draw. So, the probability that a single card drawn from a shuffled deck of cards is a spade is

$$\frac{\text{Ways to choose a spade}}{\text{Total possible ways to choose a card}} = \frac{13}{52} = \frac{\textbf{1}}{\textbf{4}}.$$

28. Two of the suits, diamonds and hearts, are red. So, there are a total of $2 \times 13 = 26$ red cards. There are 52 cards in a deck.

So, the probability of drawing a red card from a shuffled deck is

$$\frac{\text{Ways to choose a red card}}{\text{Total possible ways to choose a card}} = \frac{26}{52} = \frac{\textbf{1}}{\textbf{2}}.$$

29. Four of the cards are 7's (one from each suit), and there are 52 cards in a deck. So, the probability of drawing a 7 from a shuffled deck is

$$\frac{\text{Ways to choose a 7}}{\text{Total possible ways to choose a card}} = \frac{4}{52} = \frac{\textbf{1}}{\textbf{13}}.$$

30. Two of the cards are black Aces (A♣ and A♠), and there are 52 cards in a deck. So, the probability of drawing a black Ace from a shuffled deck is

$$\frac{\text{Ways to choose a black Ace}}{\text{Total possible ways to choose a card}} = \frac{2}{52} = \frac{\textbf{1}}{\textbf{26}}.$$

31. Twenty-six of the cards are red, including two red Kings. There are also two black Kings.

All together, there are $26 + 2 = 28$ cards that are red or Kings, including red Kings. So, the probability of drawing a card that is red or a King is

$$\frac{\text{Ways to choose a red card or King}}{\text{Total possible ways to choose a card}} = \frac{28}{52} = \frac{\textbf{7}}{\textbf{13}}.$$

32. Thirteen of the cards are spades, including 7♠. There are three other 7's: 7♣, 7♦, and 7♡.

So, there are $13 + 3 = 16$ cards that are a spade or a 7. Therefore, $52 - 16 = 36$ cards are neither a spade nor a 7. The probability of drawing a card that is neither a spade nor a 7 is

$$\frac{\text{Ways we do NOT choose a spade or 7}}{\text{Total possible ways to choose a card}} = \frac{36}{52} = \frac{\textbf{9}}{\textbf{13}}.$$

33. Fifteen of the students in the class are boys, and there are $12 + 15 = 27$ students in the class. So, the probability that a randomly selected student is a boy is

$$\frac{\text{Boys}}{\text{All students}} = \frac{15}{27} = \frac{\textbf{5}}{\textbf{9}}.$$

34. The word PROBABILITY contains 11 letters, but some are repeated. PROBABILITY contains only 9 *distinct* letters: P, R, O, B, A, I, L, T, and Y. There are 26 equally-likely letters we could choose from the English alphabet. So, the probability that a randomly selected letter of the English alphabet appears in the word PROBABILITY is

$$\frac{\text{Distinct letters in PROBABILITY}}{\text{All letters}} = \frac{\textbf{9}}{\textbf{26}}.$$

35. To create a two-digit number made only of even digits, we can choose 2, 4, 6, or 8 as the tens digit and 0, 2, 4, 6, or 8 as the ones digit. This gives a total of $4 \times 5 = 20$ two-digit numbers with only even digits.

There are 9 digits that could be the tens digit of a two-digit number, and 10 digits that could be the ones digit. So, there are $9\times10=90$ two-digit numbers.

Therefore, the probability that a randomly selected two-digit number has only even digits is

$$\frac{\text{Two-digit numbers with two even digits}}{\text{Two-digit numbers}}=\frac{20}{90}=\frac{2}{9}.$$

36. Eight primes are less than 20:
2, 3, 5, 7, 11, 13, 17, and 19.

There are 19 positive integers less than twenty.

So, the probability that a randomly selected positive integer less than 20 is prime is

$$\frac{\text{Primes less than 20}}{\text{Positive integers less than 20}}=\frac{8}{19}.$$

37. Six words in the given question have four letters: what, that, word, from, this, four.

There are a total of 15 words in the question.

So, the probability that a randomly selected word from the given question has four letters is

$$\frac{\text{Words with 4 letters}}{\text{Words in question}}=\frac{6}{15}=\frac{2}{5}.$$

38. There are 5 ways to get a sum of ten by adding a number from the top row to a number from the bottom row. To see this, we line up the numbers in the rows so each column has a sum of ten:

$$1, 2, 3, 4, 5$$
$$9, 8, 7, 6, 5$$

There are 5 choices for the top-row number and 5 choices for the bottom-row number, so there are $5\times5=25$ ways to choose one number from each row.

So, the probability that the sum of the randomly selected pair is 10 is

$$\frac{\text{Ways to make a sum of 10}}{\text{Ways to choose one number from each row}}=\frac{5}{25}=\frac{1}{5}.$$

— *or* —

No matter which number is selected from the top row, exactly 1 of the 5 numbers from the bottom row will give a sum of 10. So, the probability that the sum is 10 is $\frac{1}{5}$.

39. The factors of 48 are 1, 2, 3, 4, 6, 8, 12, 16, 24, and 48.

Among the ten factors of 48, two are odd. So, the probability that a randomly selected factor of 48 is odd is

$$\frac{\text{Odd factors of 48}}{\text{Factors of 48}}=\frac{2}{10}=\frac{1}{5}.$$

40. The factors of 77 are 1, 7, 11, and 77.

Among the four factors of 77, none are even. So, the probability that a randomly selected factor of 77 is even is

$$\frac{\text{Even factors of 77}}{\text{Factors of 77}}=\frac{0}{4}=0.$$

— *or* —

Every factor of an odd number is odd. So, every odd number has zero even factors, and it is *impossible* for a randomly selected factor of 77 to be even. The probability of any impossible event is **0**.

41. There are eight tiles numbered 8, nine tiles numbered 9, and ten tiles numbered 10. So, $8+9+10=27$ tiles have a number greater than 7.

All together, there are $1+2+3+4+5+6+7+8+9+10=55$ tiles in the bag.

So, the probability that a randomly selected tile from the bag has a number that is greater than 7 is

$$\frac{\text{Tiles numbered greater than 7}}{\text{Total tiles}}=\frac{27}{55}.$$

42. For Olivia's fraction to be less than 1, the numerator must be smaller than the denominator. We list the fractions that Olivia created, organizing by numerator:

$$\frac{1}{2}, \frac{1}{3}, \frac{1}{4}, \frac{1}{5}, \frac{2}{3}, \frac{2}{4}, \frac{2}{5}, \frac{3}{4}, \frac{3}{5}, \frac{4}{5}.$$

Of these, only $\frac{2}{3}$, $\frac{3}{4}$, $\frac{3}{5}$, and $\frac{4}{5}$ are greater than $\frac{1}{2}$.

So, the probability that the fraction Dierdra selects at random is greater than $\frac{1}{2}$ is

$$\frac{\text{Fractions Olivia writes that are greater than } \frac{1}{2}}{\text{Fractions Olivia writes}}=\frac{4}{10}=\frac{2}{5}.$$

PROBABILITY

Coin Flips   82-85

43. There are **4** equally likely outcomes:
(Penny-H, Nickel-H), (Penny-H, Nickel-T),
(Penny-T, Nickel-H), (Penny-T, Nickel-T).

44. The event described is one of the four equally likely outcomes. So, the probability that Winnie's penny lands heads and Grogg's nickel lands tails is

$$\frac{\text{Penny is heads, nickel is tails}}{\text{All outcomes of flipping two coins}}=\frac{1}{4}.$$

45. The event described is one of the four possible outcomes from our diagram: HH. So, the probability that both coins land on heads is

$$\frac{\text{Penny and nickel both land on heads}}{\text{All outcomes of flipping two coins}}=\frac{1}{4}.$$

46. There are two outcomes in which Grogg's and Winnie's coins land with the same face up: HH and TT. So, the probability that their coins land with the same face up is

$$\frac{\text{Penny and nickel show same face}}{\text{All outcomes of flipping two coins}}=\frac{2}{4}=\frac{1}{2}.$$

— *or* —

No matter what Winnie flips, Grogg's coin flip always has two possible outcomes. Only one of those outcomes will match Winnie's flip:

$$\frac{\text{Grogg's coin face matches Winnie's}}{\text{All outcomes of flipping a coin}}=\frac{1}{2}.$$

47. We can draw a diagram to show the possible outcomes of Grogg flipping two identical coins:

Coin 1	Coin 2	
heads	heads	**1.** HH
	tails	**2.** HT
tails	heads	**3.** TH
	tails	**4.** TT

Both coins landing tails is one of the four possible outcomes. So, the probability of both coins landing tails is

$$\frac{\text{Ways both coins land tails}}{\text{All outcomes of flipping two coins}}=\frac{1}{4}.$$

48. Using our diagram from the previous problem, we see that HH is one of four possible outcomes of flipping two coins. So, the probability of both coins landing heads is

$$\frac{\text{Ways both coins land heads}}{\text{All outcomes of flipping two coins}} = \frac{1}{4}.$$

Since the probability is $\frac{1}{4}$, we expect that approximately $\frac{1}{4}$ of Grogg's flips will result in both coins landing heads. One fourth of 1,000 is $\frac{1}{4} \times 1,000 = \frac{1,000}{4} = 250$.

So, we expect that both coins will land heads **between 201 and 300** times.

0-50   51-100   101-200   (201-300)   301-500   501-1,000

49. We draw the following tree diagram to represent the outcomes of the three coin flips.

Winnie's Penny	Grogg's Nickel	Alex's Dime	
heads	heads	heads	1. HHH
		tails	2. HHT
	tails	heads	3. HTH
		tails	4. HTT
tails	heads	heads	5. THH
		tails	6. THT
	tails	heads	7. TTH
		tails	8. TTT

50. Our diagram above shows the **8** possible outcomes.

51. The event described is one of the eight possible outcomes from our diagram: HHH. So, the probability that all three coins land heads is

$$\frac{\text{Ways all three coins land heads}}{\text{All outcomes of flipping three coins}} = \frac{1}{8}.$$

52. The event described is one of the eight possible outcomes from our diagram: HTH. So, the probability that Winnie and Alex both flip heads and Grogg flips tails is

$$\frac{\text{Ways Winnie \& Alex flip H, Grogg flips T}}{\text{All outcomes of flipping three coins}} = \frac{1}{8}.$$

53. There are three outcomes in our diagram in which exactly one little monster flips heads: HTT, THT, and TTH. So, the probability that exactly one little monster flips heads is

$$\frac{\text{Ways exactly 1 monster flips heads}}{\text{All outcomes of flipping three coins}} = \frac{3}{8}.$$

54. We draw a tree diagram to show all possible outcomes:

Coin 1	Coin 2	Coin 3	Coin 4	
H	H	H	H	1. HHHH
			T	2. HHHT
		T	H	3. HHTH
			T	4. HHTT
	T	H	H	5. HTHH
			T	6. HTHT
		T	H	7. HTTH
			T	8. HTTT
T	H	H	H	9. THHH
			T	10. THHT
		T	H	11. THTH
			T	12. THTT
	T	H	H	13. TTHH
			T	14. TTHT
		T	H	15. TTTH
			T	16. TTTT

The event described is one of the 16 equally likely outcomes from our tree diagram: HHHH. So, the probability that all four coins land heads is

$$\frac{\text{Ways all four land heads}}{\text{All outcomes of flipping four coins}} = \frac{1}{16}.$$

— *or* —

Each of the four coins could land heads or tails. So, there are $2 \times 2 \times 2 \times 2 = 2^4 = 16$ equally likely outcomes. The event described is one of the 16 outcomes: HHHH. So, the probability that all four coins land heads is

$$\frac{\text{Ways all four land heads}}{\text{All outcomes of flipping four coins}} = \frac{1}{16}.$$

55. There are 4 outcomes in which exactly one of the four coins lands heads: HTTT, THTT, TTHT, and TTTH. Above, we found that there are 16 possible outcomes of flipping four coins.

So, the probability that exactly one coin lands heads is

$$\frac{\text{Ways exactly 1 lands heads}}{\text{All outcomes of flipping four coins}} = \frac{4}{16} = \frac{1}{4}.$$

56. One flip of a coin does not affect the next flip. Flipping one coin four times in a row gives the same 16 possible outcomes as flipping four coins once each.

There are four outcomes in which the third and fourth flips are both heads: HHHH, HTHH, THHH, and TTHH.

So, the probability that the third and fourth flips both land heads is

$$\frac{\text{Ways last 2 flips land heads}}{\text{All outcomes of flipping four coins}} = \frac{4}{16} = \frac{1}{4}.$$

— *or* —

One flip of a coin does not affect the next flip. So, the first two flips could be anything. No matter what the first two flips are, we only need to consider the ways that the third and fourth flips could land: HH, HT, TH, or TT.

Both flips are heads in only one of these four outcomes. So, the probability that the third and fourth flips both land heads is

$$\frac{\text{Ways 2 flips land heads}}{\text{All outcomes of 2 flips}} = \frac{1}{4}.$$

57. We draw a tree diagram for each of Grogg's and Lizzie's possible outcomes.

Grogg				*Lizzie*		
Coin 1	Coin 2			Coin 1	Coin 2	
H	H	1. HH		H	H	1. HH
	T	2. HT			T	2. HT
T	H	3. TH		T	H	3. TH
	T	4. TT			T	4. TT

For each of Grogg's 4 possible outcomes, Lizzie has 4 possible outcomes. So, there are $4 \times 4 = 16$ possible outcomes when Grogg and Lizzie each flip two coins.

In each of the six following outcomes, Grogg and Lizzie flip the same number of heads:

$\underset{\text{Grogg Lizzie}}{\underline{HH}\,\underline{HH}}$, $\underset{\text{Grogg Lizzie}}{\underline{HT}\,\underline{HT}}$, $\underset{\text{Grogg Lizzie}}{\underline{HT}\,\underline{TH}}$, $\underset{\text{Grogg Lizzie}}{\underline{TH}\,\underline{HT}}$, $\underset{\text{Grogg Lizzie}}{\underline{TH}\,\underline{TH}}$, $\underset{\text{Grogg Lizzie}}{\underline{TT}\,\underline{TT}}$

So, the probability that Grogg and Lizzie flip the same number of heads is

$$\frac{\text{Ways G \& L flip same number of heads}}{\text{Ways G \& L can flip 2 coins each}} = \frac{6}{16} = \frac{3}{8}.$$

— *or* —

We can use the tree diagram from problem 54, since flipping four coins will always give 16 possible outcomes.

If we assume that Coins 1 & 2 belong to Grogg, and Coins 3 & 4 belong to Lizzie, then the following outcomes represent the cases when they each flip the same number of heads:

$$\underset{\text{Grogg Lizzie}}{\underline{HHHH}}, \underset{\text{Grogg Lizzie}}{\underline{HTHT}}, \underset{\text{Grogg Lizzie}}{\underline{HTTH}}, \underset{\text{Grogg Lizzie}}{\underline{THHT}}, \underset{\text{Grogg Lizzie}}{\underline{THTH}}, \underset{\text{Grogg Lizzie}}{\underline{TTTT}}$$

So, the probability that Grogg and Lizzie flip the same number of heads is

$$\frac{\text{Ways G \& L flip same number of heads}}{\text{All outcomes of flipping four coins}} = \frac{6}{16} = \frac{3}{8}.$$

58. Since each coin will land heads or tails, there are $2 \times 2 \times 2 = 2^3 = 8$ possible outcomes from flipping three coins. The outcomes with exactly two heads are THH, HTH, and HHT.

So, the probabillty that two coins will land heads is

$$\frac{\text{Ways two coins land H, one lands T}}{\text{All outcomes of flipping two coins}} = \frac{3}{8}.$$

Since the probability is $\frac{3}{8}$, we expect that approximately $\frac{3}{8}$ of Eve's flips will result in two coins landing heads. Three eighths of 1,000 is

$$\frac{3}{8} \times 1,000 = 3 \times \frac{1,000}{8} = 3 \times 125 = 375.$$

So, we expect exactly two of Eve's three coins to land heads **between 301 and 500** times.

0-50   51-100   101-200   201-300   (301-500)   501-1,000

PROBABILITY *Dice Rolls* 86-87

59. We fill in the chart with all possible sums of Alex's and Grogg's rolls:

Grogg's Die

	1	2	3	4	5	6
1	2	3	4	5	6	7
2	3	4	5	6	7	8
3	4	5	6	7	8	9
4	5	6	7	8	9	10
5	6	7	8	9	10	11
6	7	8	9	10	11	12

Alex's Die

60. There are six possible outcomes for Alex's roll: 1, 2, 3, 4, 5, or 6. Grogg has the same six possible outcomes. So, there are $6 \times 6 = 36$ possible combinations of rolls.

We also count **36** entries in the sum chart above.

61. In our sum chart above, we see **3** possible ways for Alex and Grogg to roll a sum of 4: (Alex 1, Grogg 3), (Alex 2, Grogg 2), and (Alex 3, Grogg 1).

62. In the previous problem, we found that there are 3 ways for Alex and Grogg to roll a sum of 4. In the problem before that, we found that there are $6 \times 6 = 36$ equally likely possible outcomes for Alex's and Grogg's rolls. So, the probability that Alex and Grogg roll a sum of 4 is

$$\frac{\text{Ways to roll a sum of 4}}{\text{All outcomes of rolling two dice}} = \frac{3}{36} = \frac{1}{12}.$$

63. There are 5 ways for Alex and Grogg to roll a sum of 8: (Alex 2, Grogg 6), (Alex 3, Grogg 5), (Alex 4, Grogg 4), (Alex 5, Grogg 3), and (Alex 6, Grogg 2).

So, the probability that Alex and Grogg roll a sum of 8 is

$$\frac{\text{Ways to roll a sum of 8}}{\text{All outcomes of rolling two dice}} = \frac{5}{36}.$$

64. a. There are 6 ways for Alex and Grogg to roll a sum of 7: (Alex 1, Grogg 6), (Alex 2, Grogg 5), (Alex 3, Grogg 4), (Alex 4, Grogg 3), (Alex 5, Grogg 2), and (Alex 6, Grogg 1).

Every other sum appears fewer than 6 times. So, **7** is the most likely sum.

b. As we saw above, there are 6 ways for Alex and Grogg to roll a sum of 7. So, the probability that Alex and Grogg roll a sum of 7 is

$$\frac{\text{Ways to roll a sum of 7}}{\text{All outcomes of rolling two dice}} = \frac{6}{36} = \frac{1}{6}.$$

65. It is impossible to roll a sum of 1 with two dice. So, to roll a sum that is less than 5, Alex and Grogg must roll a sum of 2, 3, or 4.

There is 1 way for Alex and Grogg to roll a sum of 2: (Alex 1, Grogg 1).
There are 2 ways for Alex and Grogg to roll a sum of 3: (Alex 1, Grogg 2), and (Alex 2, Grogg 1).
There are 3 ways for Alex and Grogg to roll a sum of 4: (Alex 1, Grogg 3), (Alex 2, Grogg 2), and (Alex 3, Grogg 1).

So, there are $3 + 2 + 1 = 6$ ways for Alex and Grogg to roll a sum that is less than 5.

The probability that Alex and Grogg roll a sum less than 5 is

$$\frac{\text{Ways to roll a sum} < 5}{\text{All outcomes of rolling two dice}} = \frac{6}{36} = \frac{1}{6}.$$

66. Pearl rolling one pair of dice gives the same $6 \times 6 = 36$ possible outcomes as Alex and Grogg rolling one die each. From our sum chart, we see that the probability of rolling a sum of 10 on two dice is

$$\frac{\text{Ways to roll a sum of 10}}{\text{All outcomes of rolling two dice}} = \frac{3}{36} = \frac{1}{12}.$$

Since the probability is $\frac{1}{12}$, we expect that approximately $\frac{1}{12}$ of Pearl's rolls will result in a sum of 10.

One twelfth of 1,200 is $\frac{1}{12} \times 1,200 = \frac{1,200}{12} = 100$.

So, we expect Pearl to roll a sum of 10 **between 51 and 200** times.

0-50   (51-200)   201-300   301-500   501-800   801-1,200

PROBABILITY *Geometric Probability* 88-89

67. The rectangle is split into 15 congruent regions (7 gray and 8 white). Since 7 of these 15 congruent regions are shaded, $\frac{7}{15}$ of the rectangle's area is shaded. The probability that a randomly selected point in the large rectangle is within the shaded area is $\frac{7}{15}$.

68. We split the triangle into 9 congruent regions (5 gray and 4 white) as shown:

Since 5 of these 9 congruent regions are shaded, $\frac{5}{9}$ of the triangle's area is shaded. The probability that a randomly selected point in the original triangle is within the shaded area is $\frac{5}{9}$.

69. We split the rectangle into 32 congruent triangles (16 gray and 16 white) as shown:

Since 16 of these 32 congruent regions are shaded, $\frac{16}{32}=\frac{1}{2}$ of the rectangle's area is shaded. The probability that a randomly selected point in the original rectangle is within the shaded area is $\frac{1}{2}$.

— *or* —

You may have split the rectangle into smaller congruent rectangles, as shown.

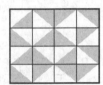

We see that exactly one half the area of every small rectangle is shaded. So, the shaded region is exactly one half the area of the total rectangle. The probability that a randomly selected point in the original rectangle is within the shaded area is $\frac{1}{2}$.

70. We split the rectangle into 30 congruent regions (22 gray and 8 white) as shown:

Since 22 of these 30 congruent regions are shaded, $\frac{22}{30}=\frac{11}{15}$ of the rectangle's area is shaded.
The probability that a randomly selected point in the original rectangle is within the shaded area is $\frac{11}{15}$.

71. We use the given measurements to determine the width of the shaded region. The width of the right unshaded region is $19-11 = 8$ inches.

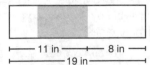

Then, since the width of the two right rectangular regions together is 15 inches and the width of the rightmost region is 8 inches, the width of the shaded center region is $15-8 = 7$ inches.

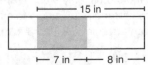

The height of each rectangular region is 4 inches.

So, the area of the large rectangle is $19\times4 = 76$ sq in. The area of the shaded rectangle is $7\times4 = 28$ sq in.

Therefore, $\frac{28}{76}$ of the rectangle is shaded. So, the probability that a randomly selected point is within the shaded area is $\frac{28}{76}=\frac{7}{19}$.

72. We first draw Paul's dartboard. Note that the bullseye can be placed anywhere within the board.

The area of the bullseye is $2\times2 = 4$ sq cm, and the area of the entire dartboard is $5\times5 = 25$ sq cm.

Therefore, $\frac{4}{25}$ of the dartboard is the bullseye. So, the probability that Paul's dart lands in the bullseye is $\frac{4}{25}$.

73. The area of the larger square is $7\times7 = 49$ sq cm. The area of the smaller square that Deanna draws is $1\times1 = 1$ sq cm. So, the area of the region inside the large square but outside the small square is $49-1 = 48$ sq cm.

Therefore, $\frac{48}{49}$ of the large square is outside the smaller square. So, the probability that Phil's randomly selected point inside the larger square is *not* inside Deanna's square is $\frac{48}{49}$.

74. The area of the shaded triangular region is equal to half of the area of the rectangle on the bottom.

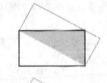

Also, we see that the longest side of the shaded triangle is equal to the long side of the tilted rectangle. When we consider that long side of the triangle as the base, the height of the triangle is equal to the short side of the tilted rectangle.

So, the area of the shaded triangular region is equal to half the area of the tilted rectangle, too!

The shaded triangular region is equal to half the area of each given rectangle. So, if the shaded area of the rectangle is a, then the unshaded area of each rectangle is also a.

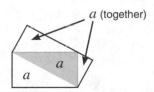

Therefore, the area of the shaded region is $\frac{1}{3}$ the area of the total figure. The probability that a randomly chosen point inside the figure is inside the shaded triangle is $\frac{1}{3}$.

75. When Alex places the disc back in the bag, there are again 20 discs numbered 1 through 20 in the bag. Alex is equally likely to pick any one of the discs.

So, the probability of drawing disc 20 is the same as it was when Alex first drew from the bag:

$$\frac{\text{Ways to pick disc 20}}{\text{Total discs in bag}} = \frac{1}{20}.$$

76. When Inid places the heart back in the deck and shuffles, the deck again contains all 52 cards. Inid is equally likely to pick any one of the cards. So, the probability of drawing a heart is the same as it was when Inid first drew a card:

$$\frac{\text{Ways to draw a heart}}{\text{Total cards in deck}} = \frac{13}{52} = \frac{1}{4}.$$

77. Lizzie's next roll is not affected by her previous roll.

There are $6 \times 6 = 36$ possible outcomes of rolling two dice, and there is only one way to get a sum of 2 with two dice ($\boxed{\cdot}\boxed{\cdot}$). So, the probability that Lizzie rolls a sum of 2 on her next roll is

$$\frac{\text{Rolls with sum of 2}}{\text{All outcomes of rolling two dice}} = \frac{1}{36}.$$

78. Cammie's coin flips are not affected by her previous flips.

So, the probability that her next three flips will all land heads is the same as the probability that she flips three heads with three flips of a coin:

$$\frac{\text{All three flips land H}}{\text{All outcomes of three coin flips}} = \frac{1}{8}.$$

79. Since Maddi eats one of the six pink candies, we know there are only 5 pink candies left in the bag. Also, there are only $5+6 = 11$ total candies left in the bag.

So, the probability that Mason picks a pink candy is

$$\frac{\text{Pink candies}}{\text{Candies in bag}} = \frac{5}{11}.$$

80. Since the first student selected was a girl, we know that there are only 9 girls left that we could select second. There are $9+15 = 24$ total students in the class that we could select second.

So, the probability that the second student selected is also a girl is

$$\frac{\text{Girls we could select second}}{\text{Students we could select second}} = \frac{9}{24} = \frac{3}{8}.$$

81. Since the bottom card of the deck is 8♢, there are only 12 diamond cards that could be the top card of the deck. Also, there are only 51 other possible cards that could be the top card of the deck.

So, the probability that the top card of the deck is a diamond is

$$\frac{\text{Diamonds that could be top card}}{\text{All possible top cards}} = \frac{12}{51} = \frac{4}{17}.$$

82. We can arrange the digits 1, 2, and 3 in $3 \times 2 \times 1 = 6$ different orders, as shown below.

$$123 \quad 213 \quad 312$$
$$132 \quad 231 \quad 321$$

Then, for each of these arrangements, there are two spots where we could place the decimal point: between the first and second digits, or between the second and third digits.

This gives us $6 \times 2 = 12$ possible numbers.

| 1.23 | <u>12.3</u> | 2.13 | <u>21.3</u> | <u>3.12</u> | <u>31.2</u> |
| 1.32 | <u>13.2</u> | 2.31 | <u>23.1</u> | <u>3.21</u> | <u>32.1</u> |

There are 8 arrangements in which the number created is greater than 3. So, the probability that a random arrangement gives a number greater than 3 is

$$\frac{\text{Numbers greater than 3}}{\text{Total possible arrangements}} = \frac{8}{12} = \frac{2}{3}.$$

83. We can arrange the four letters OPST in $4 \times 3 \times 2 \times 1 = 24$ different ways, as shown below.

OPST	<u>OPTS</u>	OSPT	OSTP	OTPS	OTSP
<u>POST</u>	<u>POTS</u>	PSOT	PSTO	PTOS	PTSO
SOPT	SOTP	<u>SPOT</u>	SPTO	<u>STOP</u>	STPO
<u>TOPS</u>	TOSP	TPOS	TPSO	TSOP	TSPO

Of these arrangements, 6 are English words: OPTS, POST, POTS, SPOT, STOP, and TOPS.

So, the probability that a random arrangement creates an English word is

$$\frac{\text{English words}}{\text{Total possible arrangements}} = \frac{6}{24} = \frac{1}{4}.$$

84. First, we look at the different groups of three digits we could use from the four digits. One way to find these groups is to choose which digit we leave out.

Group of three	Digit left out
2, 3, 4	1
1, 3, 4	2
1, 2, 4	3
1, 2, 3	4

For each of the 4 groups, there are $3 \times 2 \times 1 = 6$ ways to arrange the three digits. So, we can create $4 \times 6 = 24$ three-digit numbers from these four digits.

Next, we count how many of these numbers are divisible by both 2 and 3.

A number is only divisible by 3 if the sum of the number's digits is divisible by 3, so we compute the sum of each group of three digits.

Group of three	Sum of digits
2, 3, 4	$2+3+4 = 9$
1, 3, 4	$1+3+4 = 8$
1, 2, 4	$1+2+4 = 7$
1, 2, 3	$1+2+3 = 6$

Of these four sums, only 9 and 6 are divisible by 3. So, we can arrange 2, 3, and 4 to make a multiple of 3, and we can arrange 1, 2, and 3 to make a multiple of 3.

Any arrangement of these groups will be divisible by 3.
We list them:

123   <u>132</u>   213   231   <u>312</u>   321

<u>234</u>   243   <u>324</u>   <u>342</u>   423   <u>432</u>

A number is only divisible by 2 if the last digit of the number is even: 0, 2, 4, 6, or 8. Above, we underlined the six arrangements that are divisible by both 2 and 3.

So, the probability that the three-digit number is divisible by both 2 and 3 is

$$\frac{\text{Numbers divisible by 2 and 3}}{\text{Total possible three-digit numbers}}=\frac{6}{24}=\frac{1}{4}.$$

85. We can arrange any four people in a row in $4\times3\times2\times1=24$ different ways.
If we call the first pair of twins A and B, and the second pair of twins X and Y, then the 24 possible arrangements are shown below.

<u>ABXY</u>	<u>BAXY</u>	XABY	YABX
<u>ABYX</u>	<u>BAYX</u>	XAYB	YAXB
AXBY	BXAY	XBAY	YBAX
AXYB	BXYA	XBYA	YBXA
AYBX	BYAX	<u>XYAB</u>	<u>YXAB</u>
AYXB	BYXA	<u>XYBA</u>	<u>YXBA</u>

Each person is next to his or her twin when A&B are seated together and when X&Y are seated together. This happens in the 8 arrangements underlined above.

So, the probability that a random arrangement seats everyone next to his or her twin is

$$\frac{\text{Everyone seated with twin}}{\text{Total possible arrangements}}=\frac{8}{24}=\frac{1}{3}.$$

— *or* —

No matter who is seated in the first chair, exactly 1 of the 3 remaining people is that person's twin. So, the probability that twins occupy the first two chairs is $\frac{1}{3}$.

If one pair of twins occupies the first two chairs, the second pair of twins will also be seated together in the last two chairs. So, the probability that each person is seated with his or her twin is $\frac{1}{3}$.

86. We draw a tree diagram to represent the possible outcomes.

Of these 24 outcomes, the 14 marked with arrows indicate when Lizzie rolls a number that is greater than Winnie's.

So, the probability that Lizzie rolls a number greater than Winnie's is

$$\frac{\text{Lizzie's number} > \text{Winnie's}}{\text{Total possible outcomes of dice rolling}}=\frac{14}{24}=\frac{7}{12}.$$

87. We draw a tree diagram to represent the possible outcomes.

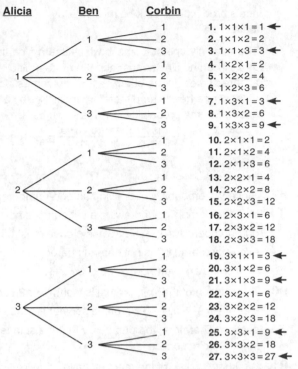

Of these 27 outcomes, the 8 marked with arrows have an odd product. The probability that the product of the three numbers is odd is

$$\frac{\text{Odd product}}{\text{Total outcomes of three spinners}}=\frac{8}{27}.$$

— *or* —

Alicia, Ben, and Corbin can each spin a 1, 2, or 3 on his or her turn. So, there are $3\times3\times3=27$ total possible outcomes.

The product of all three numbers will only be odd only if all three numbers spun are odd.

Each person can spin one of two odd numbers (1 or 3), so there are $2\times2\times2=8$ ways for all three people to spin odd numbers.

The probability that the product of the three numbers is odd is

$$\frac{\text{All spinners land on odd numbers}}{\text{Total outcomes of three spinners}}=\frac{8}{27}.$$

88. We draw a tree diagram to represent the possible outcomes.

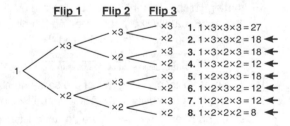

Of these 8 outcomes, the 7 marked with an arrow have an even final product. The probability that the product is even is

$$\frac{\text{Even product}}{\text{Total possible outcomes}}=\frac{7}{8}.$$

— *or* —

There are $2 \times 2 \times 2 = 2^3 = 8$ equally likely possible outcomes of three coin flips.

Joe's product is even if he multiplies by 2 (flips heads) at least once.

There is only one outcome in which Joe *never* multiplies by 2: when he flips heads (and therefore multiplies by 3) all three times. In the other $8 - 1 = 7$ outcomes, Joe flips at least one heads (and therefore multiplies by 2 at least once). So, the probability that Joe's number is even is

$$\frac{\text{Multiplies by 2 at least once}}{\text{Total possible outcomes}} = \frac{7}{8}.$$

PROBABILITY
Complements
94-95

89. Since the probability of drawing an odd-numbered cube is $\frac{1}{3}$, the probability of drawing a cube that is not odd-numbered is $1 - \frac{1}{3} = \frac{2}{3}$. Since a cube must be even-numbered if it is not odd-numbered, the probability of drawing an even-numbered cube is $\frac{2}{3}$.

90. The fraction of the spinner that is worth 1 or 3 points is $\frac{4}{9} + \frac{2}{9} = \frac{6}{9} = \frac{2}{3}$. So, $1 - \frac{2}{3} = \frac{1}{3}$ of the spinner is the 2-point section. Therefore, the probability that the spinner lands in the 2-point section is $\frac{1}{3}$.

91. We know that $\frac{7}{11}$ of the marbles have some red on them. So, the probability of selecting a marble with no red on it (all black) is $1 - \frac{7}{11} = \frac{4}{11}$.
Similarly, $\frac{9}{11}$ of the marbles have some black on them, so the probability of selecting a marble with no black on it (all red) is $1 - \frac{9}{11} = \frac{2}{11}$.
$\frac{4}{11} + \frac{2}{11} = \frac{6}{11}$ of the marbles are either all-black or all-red. Therefore, the probability of *not* getting an all-black or all-red marble is $1 - \frac{6}{11} = \frac{5}{11}$. A marble must be red with black stripes if it is not all black or all red. So, the probability of selecting a marble that is red with black stripes is $\frac{5}{11}$.

92. a. There are $6 \times 6 = 36$ possible outcomes for one roll of a pair of six-sided dice. The dice display the same number in six of these outcomes: $(1, 1)$, $(2, 2)$, $(3, 3)$, $(4, 4)$, $(5, 5)$, and $(6, 6)$. So, the probability of rolling two of the same number is

$$\frac{\text{Ways dice can match}}{\text{Total possible outcomes of rolling two dice}} = \frac{6}{36} = \frac{1}{6}.$$

— *or* —

Assume one die is red and the other is blue. No matter which number is rolled on the red die, exactly 1 of the 6 equally likely outcomes on the blue die will match the number rolled on the red die.

$$\frac{\text{Ways for blue die to match red die}}{\text{Total possible outcomes of blue die}} = \frac{1}{6}.$$

b. The probability of rolling two of the same number is $\frac{1}{6}$. So, the probability that we will *not* roll two of the same number is $1 - \frac{1}{6} = \frac{5}{6}$. If we do not roll two of the same number, then we have rolled two different numbers. The probability of rolling two different numbers is $\frac{5}{6}$.

93. a. There are $2 \times 2 \times 2 \times 2 = 16$ possible outcomes from four flips of a coin. There is only one outcome in which we flip all heads: HHHH. So, the probability of flipping four consecutive heads is

$$\frac{\text{Ways to flip four heads}}{\text{Total possible ways to flip coin four times}} = \frac{1}{16}.$$

b. If we do not flip four consecutive heads, then we have flipped at least one tails. The probability of flipping all four heads is $\frac{1}{16}$. So, the probability of flipping at least one tails is $1 - \frac{1}{16} = \frac{15}{16}$.

94. a. We subtract 99 from each number to get a list from 1 to 201. So, there are 201 numbers in this list.

Then, we count the multiples of 7 in this list. The smallest multiple of 7 that is greater than 100 is $7 \times 15 = 105$. The greatest multiple of 7 that is less than 300 is $7 \times 42 = 294$.

So, we wish to count the numbers in this list:

$$105, 112, 119, \ldots, 280, 287, 294.$$

We divide each number by 7 to get

$$15, 16, 17, \ldots, 40, 41, 42.$$

Then, subtracting 14 from each number in this list gives us a list of the whole numbers from 1 to 28. So, there are 28 multiples of 7 in the original list.

The probability that Rayad chooses a number divisible by 7 is

$$\frac{\text{Multiples of 7 from 100 to 300}}{\text{Whole numbers from 100 to 300}} = \frac{28}{201}.$$

b. In the previous problem, we found that the probability that Rayad's number is divisible by 7 is $\frac{28}{201}$.

Therefore, the probability that his chosen number is not divisible by 7 is $1 - \frac{28}{201} = \frac{173}{201}$.

PROBABILITY
Games of Chance
96-97

95. a. There are $2 \times 2 = 4$ possible outcomes from two coin flips. There are three outcomes with at least one heads: HH, HT, and TH. So, the probability that Dave gets a point is

$$\frac{\text{At least one heads on two coins}}{\text{Total possible outcomes flipping two coins}} = \frac{3}{4}.$$

b. The probability that Dave gets the first point is $\frac{3}{4}$. So, the probability that Dave will not win the first point is $1 - \frac{3}{4} = \frac{1}{4}$. Amy wins the point if Dave does not, so the probability that Amy gets the first point is $\frac{1}{4}$.

c. Amy and Dave do not have the same probability of winning a point. **Dave is more likely to win each point, and he is therefore more likely to reach 5 points first. So, this game is not fair.**

96. Each player rolls one six-sided die, so there are $6 \times 6 = 36$ possible outcomes.

Palmer wins if the product of the two numbers is odd. The product of two numbers is odd only if both numbers are odd. So, Palmer wins when he and Richard both roll odd numbers. There are 3 odd numbers they could each roll (1, 3, or 5), so there are $3 \times 3 = 9$ outcomes in which Palmer and Richard both roll odd numbers.

The probability that Palmer wins is

$$\frac{\text{Both roll odds}}{\text{Total outcomes from two dice rolls}} = \frac{9}{36} = \frac{1}{4}.$$

Richard wins if the product of the two numbers is even. Since the probability that the product is odd is $\frac{1}{4}$, the probability that the product is not odd (and therefore even) is $1 - \frac{1}{4} = \frac{3}{4}$.

So, the probability of Richard winning is $\frac{3}{4}$.

Richard and Palmer do not have the same probability of winning. Richard is more likely to win this game. So, this game is not fair.

97. There are $2 \times 2 = 4$ possible outcomes of two coin flips: HH, HT, TH, and TT.

In only one of these four outcomes (HH), Larry gets a point. The probability Larry gets a point is

$$\frac{\text{Two heads}}{\text{Total outcomes of two coin flips}} = \frac{1}{4}.$$

In two of these four outcomes (HT and TH), James gets a point. The probability that James gets a point is

$$\frac{\text{One heads, one tails}}{\text{Total outcomes of two coin flips}} = \frac{2}{4} = \frac{1}{2}.$$

Larry and James do not have the same probability of getting a point. James is more likely to win each point, and he is therefore more likely to reach 5 points first. So, this game is not fair.

98. There are $2 \times 2 \times 2 = 2^3 = 8$ possible outcomes of three coin flips, as shown below.

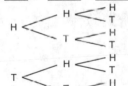

1.	Tasha flips more H
2.	Same number of H
3.	Same number of H
4.	Jason flips more H
5.	Tasha flips more H
6.	Tasha flips more H
7.	Tasha flips more H
8.	Same number of H

In four of these outcomes, Tasha flips more heads, so she gets a point. So, the probability that Tasha gets a point is

$$\frac{\text{Ways Tasha flips more heads}}{\text{Total outcomes of coin flips}} = \frac{4}{8} = \frac{1}{2}.$$

In one of these outcomes, Jason flips more heads, so he gets a point. In three of these outcomes, Jason and Tasha flip the same number of heads, so Jason gets a point. Therefore, the probability that Jason gets a point is

$$\frac{\text{Ways Jason flips more or same number of heads}}{\text{Total outcomes of coin flips}} = \frac{4}{8} = \frac{1}{2}.$$

Tasha and Jason each have the same probability of getting a point. So, neither person is more likely to get to 10 points first. This game is fair.

99. There are $3 \times 3 = 9$ possible outcomes from spinning these spinners, as shown below:

Polly	Molly	
		1. $1+1 = 2$
1	1, 2, 3	2. $1+2 = 3$
		3. $1+3 = 4$
		4. $2+1 = 3$
2	1, 2, 3	5. $2+2 = 4$
		6. $2+3 = 5$
		7. $3+1 = 4$
3	1, 2, 3	8. $3+2 = 5$
		9. $3+3 = 6$

In 4 of these outcomes, the sum is odd and Polly wins. So, the probability that Polly wins is

$$\frac{\text{Ways to spin odd sum}}{\text{Total outcomes of spinners}} = \frac{4}{9}.$$

Molly wins if the sum is even. Since the probability that the sum is odd is $\frac{4}{9}$, the probability that the sum is not odd (and therefore even) is $1 - \frac{4}{9} = \frac{5}{9}$. So, the probability that Molly wins is $\frac{5}{9}$.

Polly and Molly do not have the same probability of winning. Molly is more likely to win this game. So, this game is not fair.

100. There are $6 \times 6 = 36$ possible outcomes of two dice rolls.

There are six ways for both players to roll the same number: (1, 1), (2, 2), (3, 3), (4, 4), (5, 5), and (6, 6). So, the probability that Jeremy gets two points is

$$\frac{\text{Ways to roll same number}}{\text{Total outcomes of two dice rolls}} = \frac{6}{36} = \frac{1}{6}.$$

Since the probability that both players rolls the same number is $\frac{1}{6}$, the probability that they *do not* roll the same number (and therefore roll different numbers) is $1 - \frac{1}{6} = \frac{5}{6}$. The probability that Phyllis gets one point is $\frac{5}{6}$.

So, we expect that Jeremy wins 2 points for $\frac{1}{6}$ of the rolls, while Phyllis wins 1 point for $\frac{5}{6}$ of the rolls.

Phyllis is *five* times more likely than Jeremy to win points. However, Jeremy only receives *two* times as many points as Phyllis each time he wins points. **So, Phyllis is much more likely to reach 10 points first and win. This game is not fair.**

PROBABILITY

Pairs 98

101. With seven flavors, we can make $(7 \times 6) \div 2 = 42 \div 2 = \mathbf{21}$ pairs.

102. Three of the jellybeans are red. With three jellybeans, we can make $(3 \times 2) \div 2 = 6 \div 2 = \mathbf{3}$ pairs.

103. Four of the jellybeans are green. With four jellybeans, we can make $(4 \times 3) \div 2 = 12 \div 2 = \mathbf{6}$ pairs.

104. The probability of selecting two red jellybeans at random from the bag is

$$\frac{\text{Ways to make pair of red jellybeans}}{\text{Total ways to make pair of jellybeans}} = \frac{3}{21} = \mathbf{\frac{1}{7}}.$$

105. The probability of selecting two green jellybeans at random from the bag is

$$\frac{\text{Ways to make pair of green jellybeans}}{\text{Total ways to make pair of jellybeans}} = \frac{6}{21} = \mathbf{\frac{2}{7}}.$$

106. Any pair of jellybeans we choose from the bag will have either two red, two green, or one red and one green. So, if our jellybean pair is not two red or two green, then it is one red and one green.

In the previous problems, we found that there are 21 ways to make a pair of jellybeans, with 3 ways to make a pair of red jellybeans and 6 ways to make a pair of green jellybeans. This leaves $21-3-6=12$ ways to make a pair of one green and one red jellybean.

So, the probability of selecting one red and one green jellybean at random from the bag is

$$\frac{\text{Ways to make pair of red+green}}{\text{Total ways to make pair of jellybeans}}=\frac{12}{21}=\frac{4}{7}.$$

— *or* —

In the previous two problems, we found that the probability of selecting two red jellybeans is $\frac{1}{7}$ and the probability of selecting two green jellybeans is $\frac{2}{7}$. So, the probability of selecting two jellybeans of the same color is $\frac{1}{7}+\frac{2}{7}=\frac{3}{7}$.

So, the probability that we choose a pair of jellybeans that are *not* the same color (that is, a pair with one red and one green) is $1-\frac{3}{7}=\frac{4}{7}$.

— *or* —

There are 3 choices for a green jellybean and 4 choices for a red jellybean. This gives us $3\times4=12$ ways to make a pair with one green and one red jellybean. So, the probability of selecting one green and one red jellybean at random from the bag is

$$\frac{\text{Ways to pair green+red}}{\text{Total ways to make pair of jellybeans}}=\frac{12}{21}=\frac{4}{7}.$$

PROBABILITY
Think About It   99

There are $26!=26\times25\times24\times\cdots\times3\times2\times1$
$$=403{,}291{,}461{,}126{,}605{,}635{,}584{,}000{,}000$$
different ways to arrange all 26 letters of the alphabet in a row. We cannot possibly write out every possibility or draw a tree diagram that big, so we must think about these problems carefully!

107. The tenth letter in the arrangement could be any one of the 26 letters, with equal probability. The probability that the tenth letter in the arrangement is any particular letter (such as J) is $\frac{1}{26}$.

108. We consider one specific arrangement of the letters: QWERTYUIOP☐SDFGHJKL☐XCVBNM.

The only two letters left to place are A and Z.
We could place A in the first box and Z in the second, or we could place Z in the first box and A in the second.

Similarly, for every arrangement of the alphabet in which A comes before Z, there is another arrangement in which all other letters appear in the same spots but Z comes before A.

Therefore, the number of arrangements in which A appears before Z is equal to the number of arrangements

in which Z appears before A. The probability that A is to the left of Z for a random arrangement of the letters is $\frac{1}{2}$.

109. We consider one specific arrangement of the letters:
WS☐NH☐MJUKILOPBGTVFRCDE☐AQ.

The only three letters left to place are X, Y, and Z. There are 6 different orders in which we could place the letters in the boxes: XYZ, XZY, YXZ, YZX, ZXY, and ZYX.

Similarly, for every arrangement of the alphabet in which X, Y, and Z appear in alphabetical order (XYZ), there are five others in which all other letters appear in the same spots but X, Y, and Z appear in another, non-alphabetical order (XZY, YXZ, YZX, ZXY, or ZYX).

Therefore, the probability that X, Y, and Z appear in alphabetic order in a given arrangement of the alphabet is the same as the probability that a random arrangement of these three letters is XYZ.

Only 1 of the 6 different ways to arrange X, Y, and Z, is XYZ. So, the probability that X, Y, and Z will appear in alphabetical order is $\frac{1}{6}$.

110. We think about the arrangement in which all the letters appear in the same position they appear in the alphabet:

ABCDEFGHIJKLMNOPQRSTUVWXYZ.

For exactly one of these letters to be out of order, we would need to move exactly one letter without changing the placement of any other letters. But, to change the position of just one letter, we must move at least one other letter! So, it is impossible for exactly one letter to be in a different position.

Events that are impossible have probability **0**.

111. We think about arranging the letters in the alphabet in a circle. No matter where M is, there are two letters beside it.

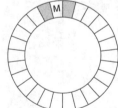

Wherever M is, there are 25 other places where N could be. In 2 of these 25 places, N will be beside M. So, the probability that M and N are next to each other when we arrange the letters in a circle is $\frac{2}{25}$.

PROBABILITY
Hats   100

112. A student is correct only if his or her guess matches the color of his or her hat.
So, a student who is guessing randomly will guess his or her hat color correctly or incorrectly, with equal probability. So, if one student guesses randomly and the other two students stay silent, then the probability that the students win is $\frac{1}{2}$.

113. If all the students guess, then they will only win if all three guesses are correct. Since each student is either correct or incorrect, with equal probability, there are $2 \times 2 \times 2 = 8$ possible outcomes for the students' guesses.

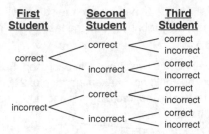

All three students guess correctly in only one of these 8 outcomes. In all the others, at least one student guesses incorrectly, so the students lose.

So, if all three students guess randomly, the probability that they will win is $\frac{1}{8}$.

114. If every student guesses, the students will only win if they are all correct.

We sketch the $2 \times 2 \times 2 = 8$ ways to assign the hats below, and label what each person says when guessing the color of the hat to his or her right.

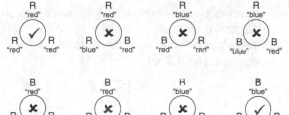

In six of these outcomes, at least one student guesses incorrectly, so the teacher wins. In the other two outcomes, all three students guess correctly. So, the probability that the students win with this strategy is $\frac{2}{8} = \frac{1}{4}$.

— *or* —

If every student guesses, the students will win only if they are all correct.

Using this strategy, each student is only correct if the hat to his or her right is the same color as his or her own. This will only happen if all hats are the same color.

Each student gets one of two hat colors, so there are $2 \times 2 \times 2 = 8$ possible hat assignments. The students win for two of these assignments (if all hats are blue or all hats are red). So, the probability that the students win using this strategy is

$$\frac{\text{Ways all hats same color}}{\text{Total number of arrangements}} = \frac{2}{8} = \frac{1}{4}.$$

115. We draw the $2 \times 2 \times 2 = 8$ possible hat assignments, and label what each person says when following Cammie's suggestion.

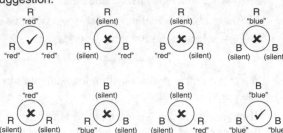

In six of these outcomes, the one student who speaks guesses incorrectly, so the teacher wins. In the two other outcomes, all three students guess correctly, so the students win. So, the probability that the students win with this strategy is $\frac{2}{8} = \frac{1}{4}$.

— *or* —

Since there are always at least two hats of the same color, at least one student will always guess. The students will win if every student who guesses is correct.

Using this strategy, a student who guesses is only correct if the hat he or she is wearing is the same color as the two other hats. This will only happen if all hats are the same color.

So, the probability that the students win with this strategy is

$$\frac{\text{Ways all hats same color}}{\text{Total number of arrangements}} = \frac{2}{8} = \frac{1}{4}.$$

116. We review our work from the previous problem. In the six outcomes in which the teacher won, only one student guessed. We notice that this student's guess would have been correct if he or she chose the *other color* from what he or she saw, rather than the same color.

So, we change just a few words in Cammie's strategy: **"Each of us looks at both hats of the other two students. If the two hats you see are the same color, guess the other color. Otherwise, stay silent."**

We compute the probability that the students win with this strategy.

Since there are always at least two hats of the same color, at least one student will always guess. So, the students will win if every student who guesses is correct.

We draw the $2 \times 2 \times 2 = 8$ possible hat placements, and label what each person says when following our new strategy.

In 2 of these outcomes, all three students guess incorrectly, so the teacher wins. In the 6 other outcomes, the one student who speaks guesses correctly.

So, the probability that the students win with this strategy is $\frac{6}{8} = \frac{3}{4}$, which is greater than $\frac{1}{2}$, as desired.

— *or* —

We compute the probability that the students win with this updated strategy.

There are $2 \times 2 \times 2 = 8$ ways to assign the hats. In 2 of these cases, all three hats match. In these cases, all three students guess the wrong color and the students lose.

In the other 6 cases, one student's hat is different than the others. In each of these 6 cases, the student with the different hat sees two hats of the same color and guesses his or her own hat color correctly. The other two students will see two different-colored hats and remain silent. So, in the 6 cases where one student's hat is different, the students win.

So, the probability that the students win with this strategy is $\frac{6}{8} = \frac{3}{4}$, which is greater than $\frac{1}{2}$, as desired.

PROBABILITY
Challenge Problems   101

117. A deck of cards contains cards of four suits. So, it is impossible to select five cards that are all from different suits.

If all five cards do not have different suits, then at least two cards have the same suit. Therefore, the probability that a group of five cards has at least two cards from the same suit is **1**.

118. No matter which suit the top card is from, 12 of the 51 remaining cards in the deck are the same suit as the top card, and $51 - 12 = 39$ remaining cards in the deck are *not* the suit of the top card. So, the probability that the second card in the deck is a different suit from the top card is $\frac{39}{51} = \frac{13}{17}$.

— *or* —

No matter which suit the top card is, 12 of the 51 remaining cards in the deck are the same suit as the top card. So, the probability that the second card in the deck is the same suit as the first is $\frac{12}{51} = \frac{4}{17}$.

The probability that the top two cards are *not* the same suit (and are therefore different suits) is $1 - \frac{4}{17} = \frac{13}{17}$.

119. We consider the different ways that James could get a sum of 6 from rolling the pair of dice: (1, 5), (2, 4), (3, 3), (4, 2), and (5, 1).

In two of these five outcomes both dice show an even number: (2, 4) and (4, 2). So, the probability that James's dice both show an even number is

$$\frac{\text{Ways to get sum of 6 with two evens}}{\text{Total ways to get sum of 6}} = \frac{2}{5}.$$

*Notice that this is a different answer than we would get by computing the probability that both dice show an even number **without** knowing the sum of the dice!*

120. No matter which color was chosen first, 4 of the 9 remaining marbles in the bag match the color of the first marble drawn. Each of the 9 marbles that remain is equally likely to be chosen last.

Since 4 of the 9 marbles are the same color as the first marble, the probability that the last marble matches the first is $\frac{4}{9}$.

121. Naymond has two choices at each hop. So, after 999 hops, he could have taken any one of 2^{999} walks. We start drawing the tree diagram and hope that we can find a pattern.

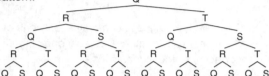

After one or three hops, Naymond can only land on R or T. Since Naymond is equally likely to hop to R or T from Q or S, he is equally likely to land at either R or T.

After two or four hops, Naymond can only land on Q or S. Since Naymond is equally likely to hop to Q or S from R or T, he is equally likely to land at either Q or S.

However long Naymond continues hopping around the diagram, this pattern must continue. We look at the diagram again, coloring vertices R and T white while leaving Q and S black:

If Naymond is at a black vertex (Q or S), then his next hop can only be to a white vertex (R or T). Similarly, if Naymond is at a white vertex then his next hop can only be to a black vertex.

Since Naymond starts at Q, after any even number of hops, Naymond will end at Q or S, with equal probability. After any odd number of hops, Naymond will end at R or T, with equal probability.

999 is odd. So, on Naymond's 999th move, he lands on point R with probability $\frac{1}{2}$.